Automatic Modeling and Fault Diagnosis of Timed Concurrent Discrete Event Systems

Automatische Modellierung und Fehlerdiagnose zeitlicher nebenläufiger ereignisdiskreter Systeme

vom Fachbereich Elektrotechnik und Informationstechnik
der Technischen Universität Kaiserslautern
zur Verleihung des akademischen Grades

Doktor der Ingenieurwissenschaften (Dr.-Ing.)

genehmigte Dissertation

von
Stefan Schneider
geb. in Landstuhl

D 386

Eingereicht am: 25.11.2014
Tag der mündlichen Prüfung: 13.02.2015
Dekan des Fachbereichs: Prof. Dr.-Ing. Hans D. Schotten

Prüfungskommision:
Vorsitzender: Prof. Dr.-Ing. habil. Wolfgang Kunz (Technische Universität Kaiserslautern)

Berichterstattende: Prof. Dr.-Ing. habil. Lothar Litz (Technische Universität Kaiserslautern)
Prof. Dr. habil. Jean-Jacques Lesage (Ecole Normale Supérieure de Cachan, Frankreich)
Prof. Dr.-Ing. habil. Ping Zhang (Technische Universität Kaiserslautern)

Bibliografische Information der Deutschen Nationalbibliothek

Die Deutsche Nationalbibliothek verzeichnet diese Publikation in der Deutschen Nationalbibliografie; detaillierte bibliografische Daten sind im Internet über http://dnb.d-nb.de abrufbar.

ISBN 978-3-8325-3981-8

Logos Verlag Berlin GmbH
Comeniushof, Gubener Str. 47,
10243 Berlin
Tel.: +49 (0)30 42 85 10 90
Fax: +49 (0)30 42 85 10 92
INTERNET: http://www.logos-verlag.de

Abstract – The productive operation of machines and facilities is of great economic importance for industrial companies. In order to achieve high productivity, unscheduled production downtimes induced by faults need to be minimized. Fault diagnosis can contribute to this aim by automatic detection and isolation of faults. In this work, a model-based fault diagnosis approach is proposed for timed concurrent Discrete Event Systems. These systems are widespread across production industry as plants controlled by programmable logical controllers. The idea of the presented fault diagnosis approach is to detect faults by comparing the observed and modeled behavior. By considering both, the logical and time behavior aspects, a wide range of different faults can be covered. The developed timed residuals allow for the isolation of the faulty system components after detecting a fault. To automatically determine the required fault diagnosis models, a timed identification algorithm has been developed. Based on observed logical and time behavior, models can be determined for large and complex industrial systems since only little system knowledge is required. In order to model concurrent system behavior, a new data-based approach has been developed that reveals concurrent behavior and determines an appropriate and optimal distributed model structure. A distributed model reflects the system concurrency and helps to reduce the number of false detections to a minimum value. The work explains the theoretical and practical aspects of the modeling and fault diagnosis approaches and gives a detailed evaluation based on a laboratory manufacturing system.

Zusammenfassung – Der produktive Betrieb von Maschinen und Anlagen ist von größter wirtschaftlicher Bedeutung für industrielle Unternehmen. Um eine hohe Produktivität zu gewährleisten, ist es notwendig, ungeplante Produktionsstillstände aufgrund von Anlagenfehlern zu minimieren. Hierzu kann Fehlerdiagnose einen wesentlichen Beitrag leisten, indem Anlagenfehler automatisch erkannt und innerhalb der Produktionsanlage lokalisiert werden. In dieser Arbeit wurde ein modellbasiertes Fehlerdiagnoseverfahren für zeitliche, nebenläufige ereignisdiskrete Systeme entwickelt. Diese Systeme sind weitverbreitet in der Industrie, üblicherweise in Form von Anlagen mit speicherprogrammierbaren Steuerungen. Bei dem entwickelten Fehlerdiagnoseverfahren wird das Verhalten einer Anlage mit einem Modell verglichen, um Fehlverhalten erkennen zu können. Dabei wird sowohl das logische, als auch das zeitliche Verhalten berücksichtigt, um eine Vielzahl von Fehlern abdecken zu können. Neuentwickelte zeitbasierte Residuen ermöglichen die anschließende Lokalisierung der defekten Systemkomponenten. Zur Gewinnung der erforderlichen Anlagenmodelle wurde ein zeitlicher Identifikationsansatz entwickelt. Basierend auf beobachtetem logischem und zeitlichem Verhalten, kann ein zur Fehlerdiagnose geeignetes Modell auch für große, komplexe Anlage ermittelt werden, da nur ein geringes Systemwissen erforderlich ist. Zur Modellierung von nebenläufigen Systemen wurde ein neuartiges, datenbasiertes Verfahren entwickelt, welches Nebenläufigkeiten in der Anlage ermittelt und geeignete, optimal verteilte Systemmodelle, erstellt. Hierdurch kann die Anzahl von Fehlalarmen bei der Diagnose auf ein Mindestmaß reduziert werden. In der vorliegenden Arbeit werden die theoretischen und praktischen Aspekte der Modellierungs- und Diagnoseverfahren aufgezeigt sowie eine detaillierte Evaluierung anhand einer Laboranwendung aus der Fertigungstechnik vorgestellt.

Acknowledgements

The work that resulted in this thesis could not have been accomplished without the support and encouragement of several people that should be mentioned here.

First of all, I would like to thank my doctoral advisor PROFESSOR LOTHAR LITZ for the opportunity to perform my research and work at his institute. I am deeply grateful for the encouragement, the trust he placed in my work, and the intellectual freedom he has given to me in these four years. Without his guidance and invaluable advice, it would not have been possible to bring this work to a successful conclusion. Besides to PROFESSOR LOTHAR LITZ, my sincere thanks go to the other members of the evaluation board: PROFESSOR JEAN-JACQUES LESAGE from the Ecole Normale Supérieure de Cachan in France, PROFESSOR PING ZHANG from the University of Kaiserslautern in Germany, and to the chairman PROFESSOR WOLFGANG KUNZ from the University of Kaiserslautern in Germany.

During my work as a research assistant at the Institute of Automatic Control I had the privilege to be part of a great team. I am grateful to all my colleagues for the rich and important discussions we had on our works and on all the other interesting topics during the coffee breaks. Thank you for you friendship, your honesty, and your humor. I deeply appreciate the time we had spent together at the institute, on our joint trips abroad, and during the numerous activities apart from work. I also would like to thank my colleagues of the LURPA in Cachan for their kindness and their inspirations that contributed to my work.

During my time at the Institute of Automatic Control, I was part of an industrial project with the company Freudenberg Vliesstoffe in Kaiserslautern. I would like to warmly thank all the people that helped to bring the project to a successful conclusion, in particular ALEXANDER BARNSTEINER for his continuous interest in my work.

My sincere thanks go to all students that contributed to my work. I especially would like to highlight the valueable works of TOBIAS CONDNE on time identification and the works of CHRISTOPH GIELEN and PAUL SALZMANN on data-based partitioning.

Many thanks to KONSTANTIN MACHLEIDT, THORSTEN RODNER, ANDREAS HAUPT, ANNA NEHRING, and ANDRÉ TELES-CARVALHO for proofreading of the manuscript and hinting at a considerable number of flaws.

Last but not least, I give the heartiest gratitude to CHRISTINE and to my parents for their love, patience, and encouragement.

Schifferstadt, Mai 2015 STEFAN SCHNEIDER

Contents

Chapter 1

Introduction

1.1 Motivation

A major economic concern of industrial companies is the productive operation of their machines and facilities. This is an essential requirement to manufacture products that are competitive in global markets. In order to ensure the productive operation of machines, their availability need to be maintained on a high level by avoiding *unscheduled operational downtimes*. Unscheduled downtimes mainly result from faults due to defective hardware components. In case of a fault, the machine needs to be restored by repairing or replacing the defective components. This requires detailed information about the location of the fault within the entire system. Fault diagnosis makes a significant contribution to this by *automatic fault detection and isolation (FDI)*. The time saved to locate a fault by using fault diagnosis provides a significant economic benefit while nevertheless additional investments for fault diagnosis implementations are comparatively low.

The FDI approach pursued in this thesis is *model-based*. The behavior of a system is compared with the modeled behavior to detect faults and to isolate the fault related hardware components. For model-based fault diagnosis, an appropriate model of the system must be determined. Various challenges arise with modeling and fault diagnosis of industrial systems. The considered systems are typically *very large* with many hardware components and perform *complex* and *time-dependent* processes. The processes often include a significant degree of *concurrency*, which requires special attention during the modeling procedure. Another major challenge is the *lack of system knowledge*. Documentation can be incomplete or became outdated over the years due to hard- and software modifications. Experts are often not available or have only limited knowledge about the considered system. Furthermore, in industrial systems many different faults can occur that are mostly unknown before they actually appear. The knowledge available is typically not sufficient, neither for modeling, nor for fault diagnosis purposes. This requires appropriate *automatic modeling* and *fault diagnosis* approaches that can operate *data-based* without relying on given knowledge.

Automatic modeling and fault diagnosis have attracted high attention in the scientific community and in industry in the last years. Various approaches have been proposed that cover a wide range of problems and applications. However, only very few works consider the field of real-world industrial systems and their mentioned, particular challenges. In this thesis, the *Discrete Event Systems (DES)* framework is used to model this class of systems. The aim is to develop new automatic modeling and fault diagnosis approaches for *timed* and *concurrent* DES. The proposed approaches cope with the challenges that come along with real-world systems and are appropriate for industrial applications. The thesis explains the theoretical and practical aspects of this work and gives a detailed evaluation using the Bosch Mechatronics System, introduced in Section 1.3.

1.2 Discrete Event Systems (DES)

The systems considered in this thesis are large-scale industrial systems. They are widely found in the industrial field, for instance communication networks, manufacturing, transportation, and power systems. To discuss automatic modeling and automatic fault diagnosis of such systems, their behavior must be formally modeled. For this purpose, the DES framework is applied, which is an approved modeling formalism that respects the event-based nature of these systems and deals with the timed behavior aspects. The framework provides a sound theory and many tools ranging from modeling and control [Brandin and Wonham, 1994] to simulation and verification [Cassandras and Lafortune, 2008].

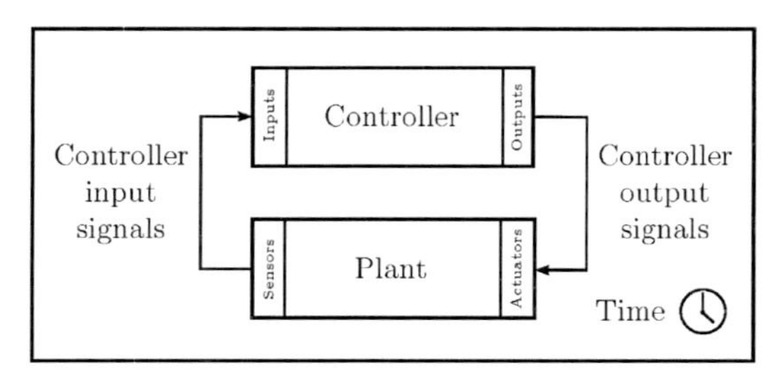

Figure 1.1: Closed-loop Discrete Event System

An industrial system is modeled as a *closed-loop DES*, as shown in Figure 1.1. It consists of a controller and a plant that are connected in a closed-loop. The controller is usually implemented as a programmable logic controller (PLC). A PLC is a multiple-input multiple-output calculation unit that automates a plant based on a defined control algorithm. The plant represents the actual facility that is to be controlled. It is equipped with sensors and actuators that interact with the PLC via the controller inputs and outputs (I/Os). The number of used controller I/Os can range from dozens up to thousands. Each sensor and actuator is connected to one controller I/O, respectively. In the considered class of systems, the signals exchanged between controller and plant have a binary value range, i.e. their value is either zero or one. The sensors measure specific process values and deliver them to the controller by using the controller input signals. The controller executes the control algorithm and computes the actuator values with respect to the measured sensor values. The actuators receive these values and actuate the process in the determined way. The communication between controller, sensors, and actuators is implemented using either hard-wired connections, bus systems, or wireless networks.

The dynamics of the DES result from the *time-dependent* interaction between controller and plant. Each step within the controlled process results from an action taken by the controller or the plant at a certain time instance. Since the controller executes a defined algorithm, its operation is considered as *deterministic*. For a given sequence of controller input signals and a given controller state, the controller computes a unique sequence of output signals in an arbitrary execution cycle. In contrast to that, the plant reacts in a *non-deterministic* way. External disturbances (friction, energy fluctuations, wear and tear, human operators, etc.) affect the operation of the plant such that its reactions can vary for different executions. As a result, the closed-loop of controller and plant must also be considered as non-deterministic. The purpose of many industrial systems is to perform *cyclic* operation processes. These systems execute permanently similar operation sequences that all lead to a required product outcome. A single sequence representing one operation cycle is called *system evolution*. Even if the product outcome is identical in different system evolutions, the actual operation sequences can differ from each other due to the non-deterministic nature of the DES. Disturbances, parametrization, and concurrency lead to a huge amount of possible sequences that may all be valid within the defined operation process.

For an efficient modeling procedure, the following formal definition of a DES given by [Cassandras and Lafortune, 2008] is introduced:

Definition 1 (Discrete Event System). "A *Discrete Event System* is a *discrete-state*, *event-driven* system, that is, its state evolution depends entirely on the occurrence of asynchronous discrete events over time."

The state-space of a DES is a finite set of discrete states. A state of a closed-loop DES is the combined state of controller and plant. No formal distinction is drawn between events caused by the controller and events caused by the plant. State transitions are observed at discrete points in time and occur instantaneously. They are triggered by events, each representing an action taken in the DES, e.g. sensing the presence of an item. These events occur *asynchronously* to the time at any instance that is typically not known in advance. Due to this property, DES are considered as *event-driven*. This is different from continuous systems, and especially from discrete-time continuous systems, since the change of the continuous state variables is closely related to the progress of time. These systems are denoted as *time-driven* systems. Since the closed-loop DES has no input signals but produces sequence of timed events, it is considered as a *timed autonomous event generator*.

A exemplary event sequence of a DES is illustrated in Figure 1.2. The discrete events e_i occur at time instances t_i. A time instance t_i is also called *time attribute* of event e_i. An event can cause the DES to change its state. In the figure, one can see that after the occurrence of event e_2 at time t_2 the system changes its state from s_2 to s_4 but after the

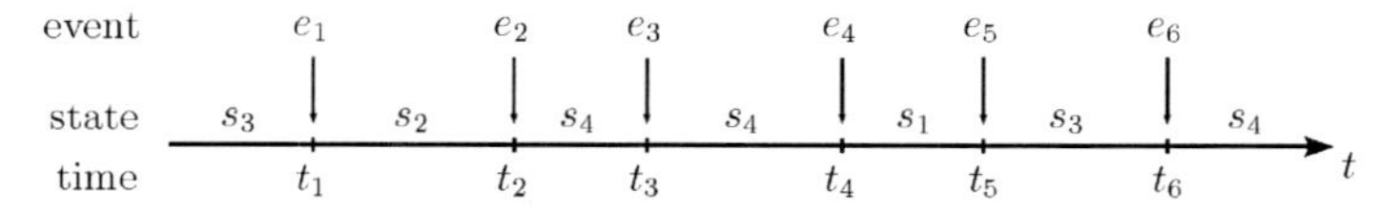

Figure 1.2: Example sequence of DES events

occurrence of event e_3 the state remains the same. The set of event sequences, a DES can perform, defines the *language* L of the system. This language represents the entire *logical behavior* of a system. In [Cassandras and Lafortune, 2008], a language L is defined as a set of words based on a given event set E. These languages are also called untimed languages in literature. Integrating the time information of the events, given by the time attributes t_i, allows to represent the *timed behavior* of the DES. This leads to the definition of the *timed language* L_t. Formal definitions of the logical and timed languages of a DES will be given in Chapter 3.

1.3 Bosch Mechatronics System (BMS)

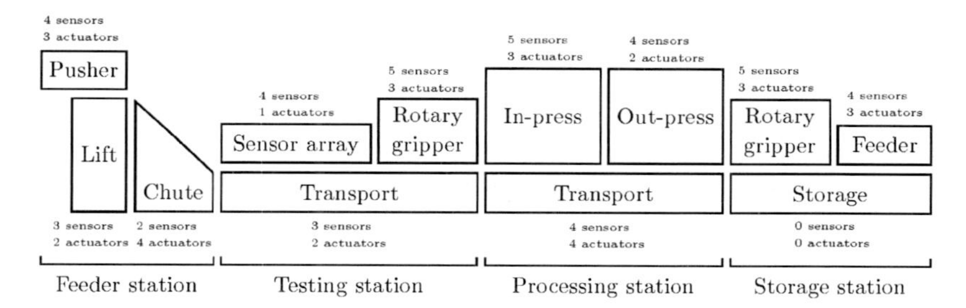

Figure 1.3: Scheme of the BMS plant

The automatic modeling and fault diagnosis approaches proposed in this work are evaluated with the *Bosch Mechatronics System (BMS)*. The BMS is a real-world laboratory manufacturing system developed by the Robert Bosch GmbH and owned by the French research institute *Laboratoire Universitaire de Recherche en Production Automatisée (LURPA)* in Cachan, France. Its original purpose is to allow technicians to study automation technology in a laboratory environment, see [Bosch Rexroth AG, 2001]. The system is equipped with a large number of industrial standard hardware components of different technologies, which can likewise be found in real-world industrial applications. The manufacturing process, executed by the BMS, is highly complex and contains a significant degree of concurrency. These properties make the system highly appropriate for evaluation purposes and allow to demonstrate the practicability of the proposed methods.

Figure 1.4: BMS plant

Figure 1.5: MODICON PLC

The plant of the BMS is composed of 4 serially arranged stations: the feeder station, the testing station, the processing station, and the storage station. Each station consists of one or more modules that perform a specific task, e.g. lifting, transporting, or pressing. The BMS plant is illustrated in Figures 1.3 and 1.4. In Figure 1.4, one can see the stations and the operation panels with the buttons for process control and emergency stop. A human machine interface is mounted on top of the plant to parametrize the process and to monitor the process variables. Additional hardware for actuation and communication is located below of the manufacturing line. The plant is equipped with 73 sensors and actuators that are required for the control of the manufacturing process. Their allocation among the plant is shown in the scheme in Figure 1.3. The 43 sensors are micro switches, reed switches, cylinder switches, capacitive, inductive, optical, and mechanical sensors. Their major purpose is to recognize the positions of machines and tools and to test and classify the workpieces. The 30 equipped actuators are mainly electronic relays and valve solenoids. They are used to switch electric drives, pneumatic valves, and electromagnets. All sensors and actuators have binary values, i.e. their value is either 0 or 1. The binary values can represent open and closed switches, present and absent workpieces, and open and closed valves, for instance. The plant is controlled using a MODICON PLC by Schneider Electric, shown in Figure 1.5. The information exchange between plant and controller is implemented using Modbus TCP[1]. Sensors and actuators are connected to the bus using remote I/O modules. Remote I/O modules are gateways that convert Modbus telegrams into binary sensor and actuator signals and vice versa. The controller periodically accesses these gateways to perform the information exchange. During one PLC cycle the sensor information is collected, the control algorithm is executed, and the actuating values are transmitted. The duration of a cycle varies between 4 and 14 milliseconds depending on computation and communication demands.

The BMS manufactures cylindrical gear wheels. Gear wheels of different materials are assembled with bearings or existing bearings are removed according to the process

[1]Modbus TCP is an open standard communication protocol for Transmission Control Protocol/Internet Protocol (TCP/IP) networks, see [Zurawski, 2005].

Parameter	Values
Number tables	$1, 2, 3$
Number workpieces each table	$1, 2, \ldots, 8$
Workpiece material	Steel, brass, synthetics
Operation mode	Insert bearing, remove bearing, advanced
Storing mode	Sort by arrival, sort by material

Table 1.1: BMS process parameters

parameters, summarized in Table 1.1. The process starts at the feeder station where the raw gear wheels are stored until they are forwarded individually to the testing station. Up to 3 pallets are placed at the feeder, each pallet equipped with 8 gear wheels. The queued wheels are either made of steel, brass, or synthetic material and have either a bearing assembled or not. They are randomly arranged on the pallets. The testing station checks the material and the bearing of each received gear wheel and stores the sensed information for the following process steps. After that, the gear wheel is given to the processing station. Depending on the sensed information and the selected operation mode, a bearing is inserted into the gear wheel, an existing bearing is removed, or the workpiece is just forwarded. Two pneumatic presses are available, one for assembling and one for disassembling. In the last stage of the process, the assembled gear wheel is given to the storage station where it is sorted into one of three storage pallets. The decision which pallet is chosen depends on the order of arrival or on its material. For an efficient use of the manufacturing line, the BMS allows to simultaneously process multiple gear wheels by concurrent operating sub-processes, e.g. the concurrent operation of the two presses; and the shared use of common process resources, e.g. the conveyors. This speeds up the overall manufacturing process and reduces the throughput time of each item.

The documentation of the BMS is a technical report [Bosch Rexroth AG, 2001]. It describes the system structure and the assembled hardware components but contains neither information about the controller algorithm, nor about the system behavior or faults. Moreover, due to several modifications and its age of more than 10 years, no more expert knowledge is available that can be used to model the behavior of the system or to perform fault diagnosis.

1.4 Contribution

The contribution of this thesis is basically three-fold: two new approaches for the automatic modeling of timed concurrent DES and a new model-based timed fault diagnosis approach are proposed. The automatic modeling approaches are identification and partitioning. The main contributions of this thesis are summarized in the following.

Identification of Timed Distributed DES Models To model the *fault-free behavior* of a timed DES, the *Timed Autonomous Automaton with Outputs (TAAO)* is introduced. A timed identification algorithm is proposed to model a TAAO *data-based* using observed fault-free output sequences of the DES. The algorithm works *automatically* and does not require any documentation or expert knowledge. According to this algorithm, the *logical* and *time behavior* of the TAAO are separately identified. While the logical behavior identification relies on an approved algorithm from literature, the identification of the time model behavior is based on *non-parametric tolerance intervals* from statistic theory. The identification parameters allow to parametrize the identification algorithm with respect to the precision and completeness of the identified model. Using identification costly manual modeling of a system can be avoided and models for even large industrial systems can be generated with reasonable efforts. The timed identification approach for monolithic models is extended to identify a *timed distributed model*. This facilitates the identification of complete models for systems with a high degree of concurrency. Due to the precision and completeness guarantees, which can be given for the identified models, these models can be appropriately used for model-based fault diagnosis purposes.

Partitioning of DES Models A new method for the automatic partitioning of models for concurrent DES is proposed. The aim of partitioning is to *automatically* determine the distributed subsystem structure of a DES in order to model each subsystem by a partial TAAO. As identification, partitioning works *data-based* using the observed fault-free output sequences and does not rely on documentation or expert knowledge. The proposed approaches in this thesis are *causal partitioning*, *optimal partitioning*, and *partition synthesis*. With causal partitioning, subsystems of the DES are determined by recovering causal relationships between hardware components. In the second approach, optimal partitioning, subsystems are determined by considering an optimization problem. The applied optimization criteria ensure that the generated partial models meet desired properties. The determined feasible subsystems are used to compose the *I/O-partition* of the closed-loop DES. This is done during partition synthesis. Based on the automatically determined I/O-partitions, a timed distributed model of the DES can be identified.

Fault Detection and Isolation using Timed Distributed DES Models A model-based fault diagnosis framework is proposed to perform *on-line fault detection and isolation* of *logical* and *time related faults*. The applied model is either a monolithic TAAO or a timed distributed model, which can be generated using the proposed timed identification approach. Timed FDI relies on a timed state estimation and on the determination of important time constraints by the modeled behavior. The timed fault detection algorithm is applied to compare the observed and modeled behavior of the DES to detect logical and time related misbehavior. Specific *timed residuals* for DES are developed in this thesis in order to isolate time fault symptoms. By use of the residuals, a set of fault candidates can be concluded that are related to a detected fault. Extensions for the FDI procedures are introduced in order to perform *timed distributed FDI* using timed distributed models. This allows for a low number of false detections with concurrent operating systems. It will be presented how local FDI is performed for each partial TAAO of the distributed model and how local results can be combined to the global FDI result.

1.5 Organization

After this introduction, the challenges of automatic modeling and fault diagnosis for the considered class of DES follow in Chapter 2. It starts with the definition of basic fault diagnosis terms and concepts that are necessary to discuss fault diagnosis. After that, the challenges for a model-based FDI framework are introduced. These challenges have again several implications for the applied models and the developed automatic modeling approaches. In Chapter 3, the proposed approach for identification of timed distributed DES models is presented. Initially, the timed behavior modeling of a DES is addressed. The timed languages, untimed languages, and time spans are defined and explained in the context of original, observed, and modeled behavior. Next, the timed identification algorithm is introduced and illustrated by means of an example. The algorithm is initially discussed for a monolithic timed automaton and then extended to a distributed modeling framework. The precision and completeness properties of an identified automaton are analyzed and the meaning of the identification parameters is discussed. In order to show the capabilities of monolithic and distributed identification, the proposed identification algorithm is evaluated with the BMS.

In Chapter 4, the partitioning of DES models is considered. After giving an overview about the partitioning problem, the three partitioning procedures causal partitioning, optimal partitioning, and partition synthesis are presented. It is shown how causal relations between hardware components can be determined by causal partitioning and how to use them for partitioning of a DES model. In the section of optimal partitioning, the optimization problem is formulated and a modified stochastic hill climbing algorithm is presented to solve the presented optimization problem. Then, partition synthesis is considered in terms of a second, consecutive optimization problem. The presented partitioning approaches are applied to the BMS.

In Chapter 5, the new approach for fault detection and isolation using a timed distributed model is presented. Evaluation, timed fault detection, and timed fault isolation procedures are introduced for monolithic and for timed distributed models. The evaluation consists of a timed state estimation algorithm and a procedure for determining time constraints relevant for FDI. After presenting the evaluation procedures, the timed fault detection and isolation methodology is considered. For timed fault isolation, special timed residuals are developed that follow the idea of DES residuals, which were originally presented in [Roth, 2010]. After this discussion, the proposed FDI approach is evaluated with the BMS. It follows a description of the FDI implementation, a presentation of fault scenarios, and a validation of identified DES models.

In Chapter 6, an analysis of related literature is given. The review starts with a discussion of models that are commonly used for identification and fault diagnosis of DES. Then, existing identification approaches are presented and compared. Works that address the identification of concurrent system are considered in the next section. In the remaining part of the review, an overview about existing fault diagnosis approaches that use logical and timed models is given. The thesis is finalized with the conclusion, suggestions for further works, and the extended summary in German language.

Chapter 2

Automatic Modeling and Fault Diagnosis Challenges

2.1 Fault Diagnosis of DES

2.1.1 Terms

To discuss the challenges of fault diagnosis for DES, it is necessary to define the basic terms. First, a distinction is drawn between *fault* and *failure*. In the international standard [ISO/IEC 2382-14:1997, 1997] a fault is defined as:

Definition 2 (Fault). "An abnormal condition that may cause a reduction in, or loss of, the capability of a functional unit to perform a required function."

The term fault refers to the state of a system or hardware component that is entered upon the occurrence of a defect. However, a faulty system can still be able to perform its required function. One can think of redundant hardware components, in which the defect of one does not affect the required functioning of the system. The standard [IEEE Std 100-1996, 1997] distinguishes between *intermittent* and *permanent* faults. Intermittent faults are temporary defects caused by degradation or inadequate system design. They can spontaneously disappear after their occurrence such that the system is again in its fault-free state. In contrast to that, permanent faults remain stable after their occurrence and can only be removed by adequate maintenance services. In this thesis, no distinction is drawn between these two fault types. If a fault leads a system to be unable to perform its required function, then the system is failed. The definition of failure is given in [ISO/IEC 2382-14:1997, 1997] as:

Definition 3 (Failure). "The termination of the ability of a functional unit to perform a required function."

In contrast to the definition of fault, a failure is an event. The cause for a system failure can be more than one faulty hardware component.

The diagram in Figure 2.1 gives a closer look at the relationship between faults and failures in a DES. In the figure, the DES is considered as a hierarchical composition of

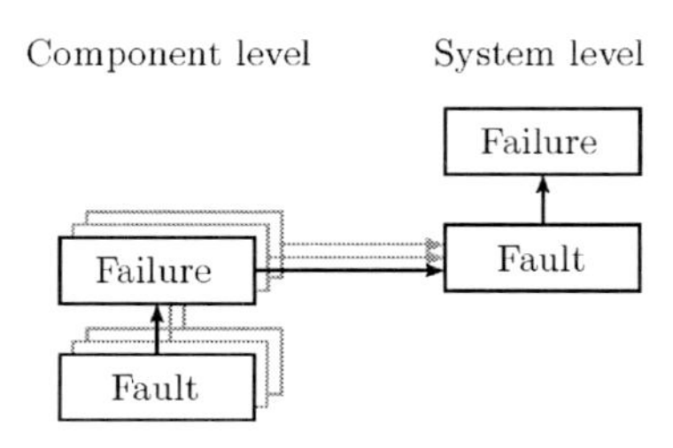

Figure 2.1: Relationship between fault and failure in a DES according to [ISO/IEC 2382-14:1997, 1997]

two levels: the component level and the system level. The required function of the DES is described on the system level. If the DES is in the fault state, it may fail or not, depending on the nature of the fault. The component level describes the conditions of hardware components of the DES. Depending on the degree of abstraction, a component can represent a single sensor or a subsystem of the DES, for instance. A faulty hardware component can fail such that it cannot fulfill its task within the process any more. In this case, the failed hardware component forces the system to enter the fault state. A fault on the system level requires that one or more hardware components have failed.

The occurrence of a fault is in general not directly observable. However, faults in industrial systems force the system to behave in a way that is somehow abnormal. An observable abnormal behavior is denoted as *fault symptom* and is defined in the standard [IEEE Std 100-1996, 1997] as:

Definition 4 (Fault symptom). "A measurable or visible abnormality in equipment parameters."

A fault can only be recognized by a fault diagnosis system if it leads to an abnormal system behavior. If a system behaves the same in a faulty as in the fault-free state, then the fault cannot be recognized.

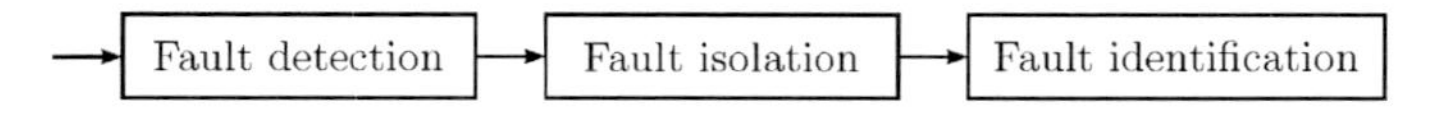

Figure 2.2: Fault diagnosis tasks

The task of fault diagnosis is the recognition and the interpretation of fault symptoms. The basic diagnosis tasks are *fault detection*, *fault isolation*, and *fault identification*, see [Chen and Patton, 1999]. The procedure is illustrated in Figure 2.2. First, with fault detection a decision is made whether the DES is fault-free or faulty. This decision depends on whether any fault symptoms are recognized. In case of a detected fault, the next step is the fault isolation task. During fault isolation, the location of the fault is determined by interpreting the observed fault symptoms, i.e. which sensors and actuators are fault candidates. Fault candidates are those hardware components whose misbehavior can explain the detected fault. Finally, the size and the nature of the isolated fault is determined by fault identification to arrange for the required repairs.

2.1.2 Concepts

To avoid unscheduled operation downtimes due to system failures, *preventive* and *corrective maintenance* strategies are applied [Birolini, 2010]. The idea of preventive maintenance is to replace system components before they actually become faulty. This serves to prevent from failures that are mainly a result of wear and tear [Birolini, 2010]. However, due to the stochastic nature of faults, it is not possible to avoid failures completely. In case a system fails, corrective maintenance is applied. The task is to repair or to replace failed system components in order to restore the required function of the system. Another concept that makes use of diagnostic information besides corrective maintenance is *fault-tolerant control*. The idea is to modify the control law after the occurrence of a fault such that the effects of the faulty component on the system are limited or even removed. This can be achieved by using redundancies as presented in [Blanke et al., 2006] for continuous systems. In [Nke and Lunze, 2011], fault-tolerant control is ported to DES and applied to an manufacturing system. The authors show how a controller can be on-line reconfigured in order to cope with sensor and process faults. To apply this approach, precise information about the occurrence and the location of the fault need to be available, which can be determined by appropriate fault diagnosis concepts.

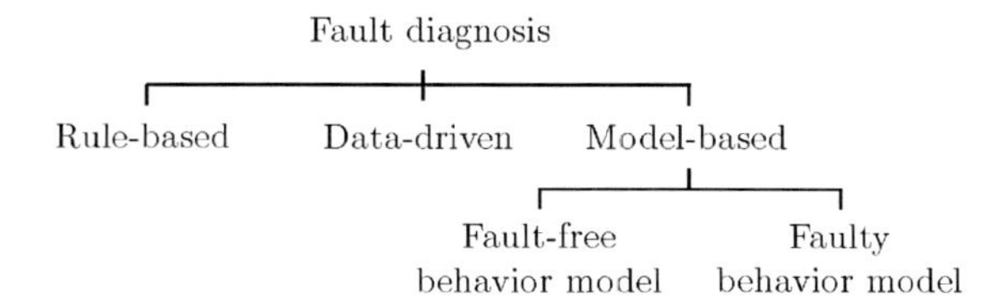

Figure 2.3: Fault diagnosis concepts [Papadopoulos and McDermid, 2001]

Before hardware components can be repaired or a control law can be adapted, an occurred fault has to be located by fault diagnosis. Following the classification of [Papadopoulos and McDermid, 2001], the three basic fault diagnosis concepts are: *Rule-based*, *data-driven*, and *model-based* fault diagnosis, as shown in Figure 2.3. Fault diagnosis using a rule-based expert system relies on a collection of rules that describe the relationships between faults (cause) and observed fault symptoms (effects) [Papadopoulos and McDermid, 2001]. The knowledge-base is learned by experts which are highly experienced in the considered field. All rules are formulated in terms of conditional statements (IF-THEN), e.g.: "IF e_2 is observed before e_1 THEN *switch* is faulty", with e_1 and e_2 are events risen by the DES. The fact "e_2 is observed before e_1" is an observed fault symptom that is obtained by observing the events of the system. The cause "*switch* is faulty" is a deduced conclusion that explains the observation. In general, the conclusion of a rule can be again the cause for another rule. A chain of rules is executed in order to find the most likely reason for the fault. Expert FDI systems are successfully applied to systems when the number of rules is small. For large and complex industrial systems, the rule base grows huge and tends to be inconsistent and incomplete with a lack of portability and maintainability [Papadopoulos and McDermid, 2001]. Due to these reasons, rule-based expert systems are less suitable for the fault diagnosis of industrial DES.

Data-driven fault diagnosis approaches make use of the trend information that are contained in sensor and actuator signals. A simple example for signal-based fault diagnosis of an electric motor is to monitor the motor current and raise an alarm when it exceeds a given threshold. To obtain this information, no model of the process is needed but the threshold must be known. Advanced approaches use spectral analysis in order to extract more information, e.g. the signal frequency. This is especially of interest with rotating machines [Lunze, 2012]. The concept of qualitative trend analysis mentioned in [Dash and Venkatasubramanian, 2000] is another data-driven fault diagnosis approach. A predefined set of trend patterns, called primitives, constitute the language that is used to represent sensor trends. After extracting the sensor trends from the data, they are matched against a knowledge-base with known trends. If the observed trend cannot be reproduced, then a fault is diagnosed. Data-driven methods are well-adapted in continuous systems for which trends in signals over the time are available. The strength of these methods is to compare observed trends with a historical reference and to deduce diagnostic information from this. Typically, thresholds for process values exist that are used to detect faults in the system [Lunze, 2012]. This allows to detect faults but provides only little information where it may be located. Trends in DES exist in the form of event and state sequences. The basic comparison of current and former state sequences does not allow for diagnostic statements since the sequences generated by the system are not unique and deviations are permitted. Also constraint monitoring is not applicable for DES due to the binary value range of the sensors and actuators. In summary, data-driven methods are a powerful diagnostic tool for continuous systems but less suitable for the diagnosis of DES.

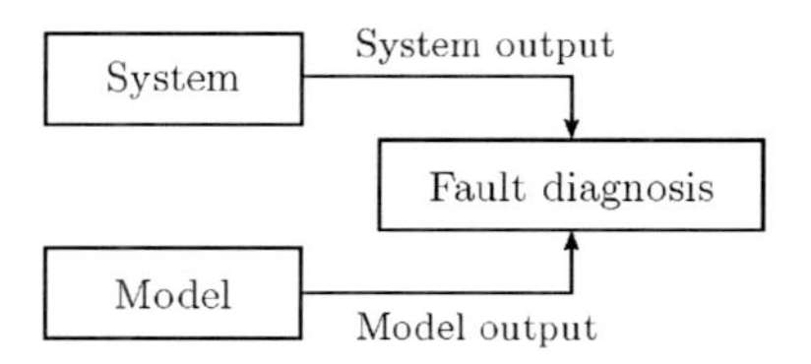

Figure 2.4: Model-based fault diagnosis concept

Approaches that use a model of the considered system for fault diagnosis are called *model-based*. Model-based fault diagnosis follows the principle of *consistency testing* [Blanke et al., 2006], i.e. it is checked whether observations of the system behavior are consistent with the modeled behavior. This concept is illustrated in Figure 2.4. Model-based fault diagnosis approaches can be basically divided into approaches that use *fault-free models* and into approaches that use *fault models* of the system. While fault-free models describe the fault-free system behavior only, fault models contain additional information about the DES behavior for specific fault scenarios. Faults can be detected by comparing the observed and the modeled fault-free behavior. If there is a deviation, the behavior of the system cannot be considered as fault-free any longer, hence the system is in a fault condition. If information about faults are modeled, it can be checked whether the observed faulty behavior is consistent with the modeled fault scenarios. In that way, faults can be unambiguously isolated.

In this work, the model-based diagnosis concept is chosen for fault diagnosis. In contrast to rule-based expert systems and data-driven approaches, the model-based diagnosis concept is well suited for industrial systems. The model formalisms provided by the DES framework are powerful tools to describe the behavior of such systems in a compact and clear way. Their natural dynamic behavior is represented by the state- and event-based view of these formalisms. However, the major challenge using model-based approaches is the generation of an appropriate system model, especially for industrial systems. The model and the modeling procedure have to satisfy several requirements to be suitable for fault diagnosis purposes. In the following section, these requirements are discussed including the question of whether fault-free or faulty system models have to be used.

2.2 Model-based Fault Diagnosis Challenges

A model-based fault diagnosis approach for industrial systems has to satisfy several requirements. They come along with the considered class of systems or are placed by users of the diagnosis system. A basic requirement is to perform fault diagnosis *automatically* without requiring manual interventions. The aim is to diagnose systems that cannot be handled manually due to their size and complexity and to relieve the system operators from time-consuming monitoring tasks. A further advantage of automatic diagnosis is its *formalized* and *standardized* proceeding for fault isolation. This provides comprehensive and transparent diagnosis results, which do not rely on the diagnosis abilities of the maintenance crews. In the following, an enumeration of the most relevant issues for automatic model-based fault diagnosis of timed DES is given.

Fault detection and isolation Fault diagnosis for DES is considered in this work as *fault detection and isolation* of the fault related hardware components. The practical interest is mainly to detect and to isolate faults that lead a system to fail. Fault identification is not considered as a part of automatic fault diagnosis. In industrial companies, this task is typically carried out manually by a maintenance crew after fault isolation. In case of a fault, the FDI system has to detect it and to provide a set of fault candidates that are related to the fault, e.g. "limit switch xy is faulty". In order to cover a wide range of different faults, the diagnosis approach has to deal with *logical* and *time fault symptoms*. A logical fault symptom is, for instance, the absence of a sensor signal that leads to an undesired system behavior. Time fault symptoms refer to a timed misbehavior of a hardware component. A worn-out actuator may still be able to perform its function but due to the decreased efficiency its reaction may be slower than required.

Minimal missed and false detections Fault diagnosis should avoid missed and false detections. A formal representation of this is to consider fault detection as a hypothesis test. The task is either to accept or to reject the hypothesis H_0 : "The system is in the fault state". In the case the system is faulty and H_0 is accepted by the diagnosis, a truly occurred fault is detected. This is the desired reaction of a fault diagnosis system. However, it may happen that a fault occurred but the diagnosis rejects H_0. Then, the diagnosis system made a wrong decision and the fault is not detected. This is a *missed detection*. The other desired reaction of a fault diagnosis system is when the system is operating fault-free and the diagnosis rejects the hypothesis H_0. Then, the fault-free state of the system is correctly determined and no alarm is raised. A wrong decision would be made if the diagnosis accepts H_0, although no fault is present in the system. This case refers to a *false detection*. A FDI system is supposed to have a minimal number of missed and false detections. Missed detections should be avoided in order to obtain a high usability. An FDI system that misses many faults is obviously useless. The number of false detections should also be minimal to avoid unnecessary alarms and to improve the user's confidence in the diagnosis system. However, the experience with FDI systems shows that missed and false detections can be minimized but typically not completely avoided. The reasons therefore are modeling uncertainties and disturbed observations mostly. In addition to that, the rates of missed and false detections are interdependent. Reducing one rate by appropriate methods often increases the other one. An important practical issue of developing a fault diagnosis system is to choose an appropriate *trade-off* between both.

Passive FDI It is required for industrial systems that a fault diagnosis system does not affect or interrupt the operating process. In [Sampath et al., 1998] a diagnosis approach is proposed in which the diagnosis system actively affects the operation of a DES to improve the diagnosis result. This is not permitted for the considered industrial systems, since an active intervention can lead to undesired production results or even operation stops.

Online FDI Fault diagnosis has to be performed online during system operation to provide information about the fault as early as possible after its occurrence. This requires efficient algorithms for monitoring and fault diagnosis in order to execute these tasks in real-time.

Economical implementation The costs for implementing a model-based fault diagnosis solution have to be low compared to the economic benefit. This includes the costs for engineering and especially for modeling of the considered system.

2.3 Modeling Challenges

The key issue in the context of model-based fault diagnosis is the system model. Modeling a technical system is basically not a unique procedure. Depending on the purpose of the model and the degree of abstraction, different models may describe the same system.

Modeling can be either performed *manually* or *automatically*. Manual behavior modeling of *large* and *complex* DES is a very time consuming procedure due to the large amount of behavior a DES can perform. It requires detailed system knowledge that must be available in form of documentation or given by experts. However, over the years, industrial systems are typically modified for several times, intentionally for optimization reasons and unintentionally due to wear and tear. Documentation and expert knowledge becomes incomplete and outdated and is no longer sufficient for modeling purposes. In order to overcome these problems, *automatic* modeling methods can be applied that work *data-based* using observations of the system behavior. This allows to model even large and complex systems in an economical way. The modeling approaches considered in this thesis are *identification* and *partitioning*. Since the generated models are used for fault diagnosis purposes, they have to meet several requirements that have to be ensured by the modeling approaches. These requirements are discussed in the following.

Timed models A system model must be capable of representing the essential information that is required for the intended application. In the context of timed fault diagnosis of timed DES, the relevant information are the *logical* and the *time system behavior*. While the logical behavior is necessary to detect and isolate logical faults, the time behavior is required for time related faults. Furthermore, an applied model must be able to represent the non-deterministic nature of a closed-loop DES.

Fault-free behavior Model-based FDI approaches use models that represent either the fault-free or faulty behavior of the considered systems. For industrial DES, the fault-free modeling concept has to be chosen due to several reasons. To automatically generate models of the faulty behavior, information about the faults must be available, either in form of documentation and expert knowledge or based on observations. Documentation and expert knowledge do usually not exist for faulty behavior of industrial systems. The identification of faults is also not practical in general, since faults typically occur only very sporadic and randomly. Consequently, models that are used for FDI have to represent the *fault-free behavior* of the DES that can be *passively* observed during its fault-free operation. In this thesis, the term fault-free behavior is used to describe the behavior of a DES that is *accepted* by a system operator rather than the 'idealized' behavior. System operators typically evaluate the functioning of the system by the specified product quantity and quality. As long as the DES provides the desired product, it is considered as operating fault-free and the resulting event sequences are accepted. However, it cannot be excluded that an accepted sequence contains a misbehavior of the system that does not affect the system functioning. This misbehavior cannot be recognized by a system operator. Accepting these sequences does not restrict the capabilities of the proposed fault diagnosis approach since these faults do not prevent the system to perform its required function. The terms fault-free behavior and accepted behavior are synonymously used in the remaining of this thesis.

Precise and complete models A major demand on fault diagnosis is to have a minimum number of missed and false detections. This requires models that are *precise* and *complete*.[2] To discuss these properties in the context of timed models, the timed languages L_t, represented by the DES, and the applied model have to be considered. As outlined in Section 1.2, the language of a system or model can be defined in different ways. Nevertheless, the following general considerations hold for any logical, timed, or stochastic timed language definition.[3]

An automatically generated timed model that is used for fault diagnosis should model the *original timed language* $L_{Orig,t}$ of the DES as accurately as possible. This language represents the fault-free timed behavior of the system. For automatic modeling, a set of system observations is available that form the *observed timed language* $L_{Obs,t}$. Since the observation horizon is bounded, the observed timed language is in general a more or less complete subset of the original timed language $L_{Obs,t} \subseteq L_{Orig,t}$. Automatic modeling uses $L_{Obs,t}$ to build a model that is able to reproduce the *modeled timed language* $L_{Mod,t}$. The aim is to model the DES in such a way that the modeled timed language is equal to the original timed language $L_{Mod,t} = L_{Orig,t}$. This is achieved, if the system is completely observed $L_{Obs,t} = L_{Orig,t}$ and an applied identification algorithm generates a model that exactly reproduces the observed language $L_{Mod,t} = L_{Obs,t}$. However, since the original timed language of a DES is unknown, it cannot be guaranteed that it is completely observed for a given observation horizon $L_{Obs,t} \subset L_{Orig,t}$. This leads in general to a model whose language does not perfectly match with the original one $L_{Mod,t} \approx L_{Orig,t}$.

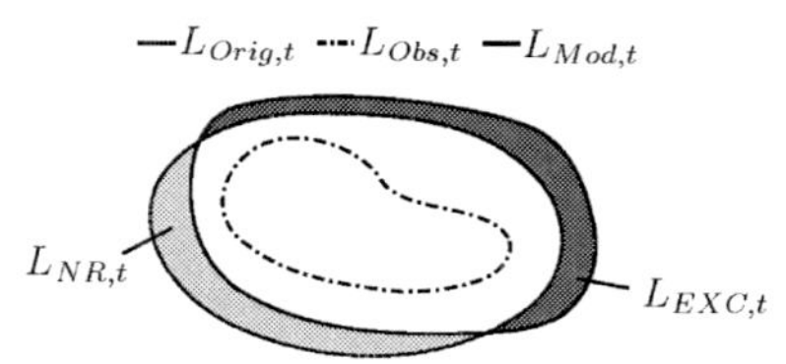

Figure 2.5: Relationships between languages

In Figure 2.5, the relationships between the original, observed, and modeled timed languages are shown for the general case. The original timed language $L_{Orig,t}$ is the area enclosed by the regularly dashed line. Within this area, the observed timed language $L_{Obs,t}$ is located. The language obtained with the model $L_{Mod,t}$ is illustrated by the area based on the solid line. It is reasonable to assume that the observed timed language is contained in the modeled one $L_{Mod,t} \supseteq L_{Obs,t}$ since an automatically generated model should at least be able to reproduce the observed timed behavior. Automatic modeling approaches typically build models that represent a language that is a proper superset of the observed one. One can see in Figure 2.5 that the area of the modeled timed language is significantly larger

[2]Precision and completeness are introduced in [Medeiros et al., 2007] in the context of system identification. They should not be confused with precision and recall used in classification and statistical testing.

[3]See Chapter 6 for a review on logical, timed, and stochastic timed languages.

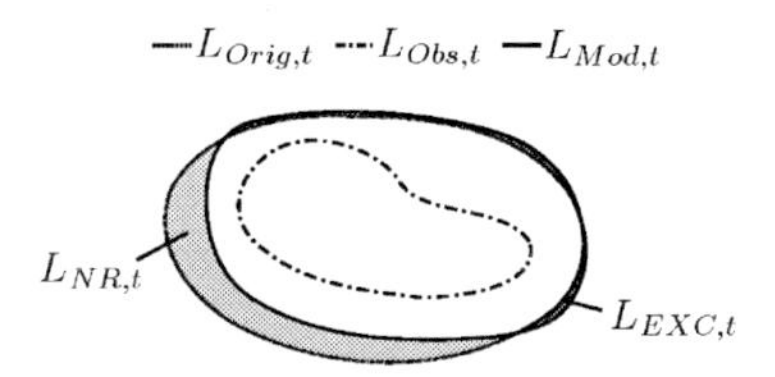

Figure 2.6: Languages relationship of an approximately precise model

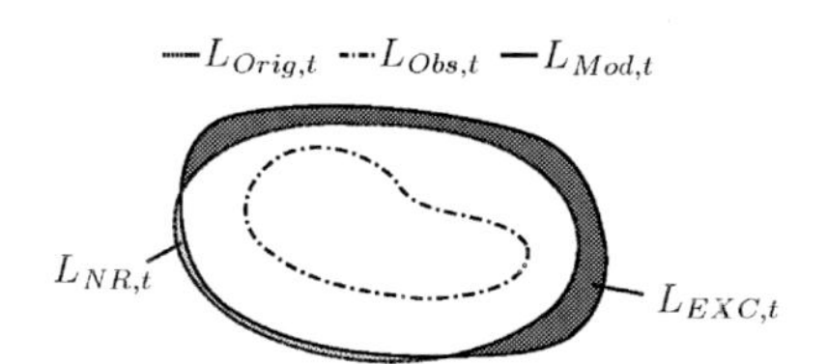

Figure 2.7: Languages relationship of an approximately complete model

than the area of the observed timed language. If this additional language is not part of the original one, it belongs to the *exceeding timed language* $L_{EXC,t} = L_{Mod,t} \setminus L_{Orig,t}$. This language represents behavior that can be reproduced by the model but it is not a part of the original fault-free timed system behavior. If a model has no exceeding timed language, it is called *precise*. The exceeding timed language $L_{EXC,t}$ of a FDI model has a significant impact on fault detection. Exceeding behavior may correspond to DES behavior that is generated in case of a fault. This means that faulty behavior of the DES can unintendedly be modeled by means of the exceeding language $L_{EXC,t}$. Such faulty behavior, generated by the DES, cannot be detected by a fault detection algorithm since observed and modeled behavior are consistent. The result is a *missed detection*.

If the original timed language of a DES is not completely observed $L_{Obs,t} \subset L_{Orig,t}$, it cannot be ensured that an automatically modeled timed language contains the complete original timed language $L_{Mod,t} \not\supseteq L_{Orig,t}$ of the DES. This leads to the *non-reproducible timed language* $L_{NR,t} = L_{Orig,t} \setminus L_{Mod,t}$. It contains those elements of the original timed language that are not part of the modeled one. If a model has no non-reproducible timed language, it is called *complete*. As $L_{EXC,t}$, the non-reproducible timed language $L_{NR,t}$ also has an impact on fault detection. If a non-reproducible timed behavior is observed, the DES is basically in its fault-free condition and performs an accepted operation. The model, however, is not able to reproduce this behavior since it is not part of the modeled language. As a result, the fault diagnosis detects an inconsistency between the observed and modeled behavior and concludes that the system must be faulty. These *false detections* can be avoided if complete models are used for FDI.

To avoid missed and false detections, the applied models are supposed to be precise and complete. However, since the original timed language of a system is unknown and the observation horizon is limited, it cannot be guaranteed in general that an automatically generated model contains no exceeding and no non-reproducible timed language. The aim of automatic modeling is rather to generate models that are as precise and complete as possible based on the given observed outputs sequences. In Figures 2.6 and 2.7, the language relationships of an approximately precise and complete model based on the same original and observed timed languages are illustrated, respectively.

Chapter 3

Identification of Timed Distributed DES Models

3.1 Preliminaries

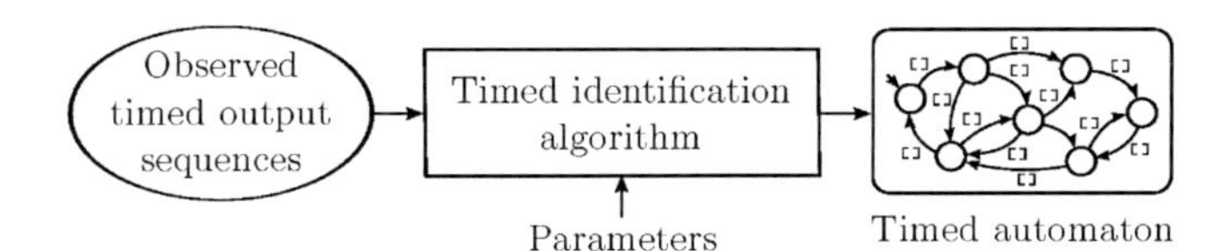

Figure 3.1: Overview timed identification approach

In this chapter, the *automatic data-based* modeling of timed concurrent DES is addressed. The aim is to build timed automata models that are appropriate for timed FDI purposes. In [Cassandras and Lafortune, 2008], the behavior of a system is modeled by means of logical, timed, and stochastic timed languages. A model is supposed to reproduce, or to 'speak' the language of the considered system. The two commonly used modeling concepts able to represent languages of DES are *automata* and *Petri nets* [Cassandras and Lafortune, 2008]. Both frameworks are well suited to model DES due to their event- and state-based formalism. Many works have been proposed in the scientific community that deal with the automatic data-based generation of these models, called *model identification*. [Booth, 1967], [Klein, 2005], and [Roth, 2010] address the identification of automata models while Petri Nets are identified in [Meda-Campaña and López-Mellado, 2005] and [Cabasino et al., 2007]. The models identified in these works represent logical languages. Considerably less works deal with the identification of timed models, in which the timed behavior of a system is modeled in addition to the logical one. Works in this field are [Grinchtein et al., 2005] and [Supavatanakul et al., 2006] for timed automata identification and [Basile et al., 2011] for timed Petri nets. A comprehensive discussion of literature works and examples are given in Chapter 6.

The aim in this thesis is to identify a timed automaton model that represents the *fault-free* timed language of a closed-loop DES. The identification concept is illustrated in Figure 3.1. The data used for identification are observed timed output sequences of the fault-free DES behavior. The proposed timed identification algorithm is based on the approved logical identification algorithm after [Roth, 2010]. In order to deal with concurrent behavior, the timed identification approach is extended to identify a *distributed* model. Partial models are built for each subsystem such that their combined behavior represents the global concurrent system behavior.

3.2 Timed Modeling

3.2.1 Internal and External Behavior

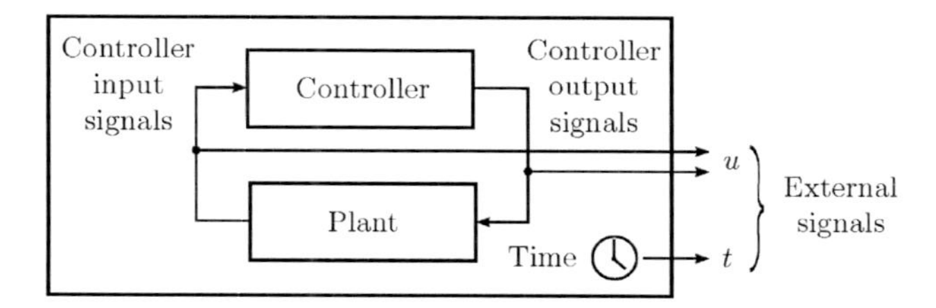

Figure 3.2: Passive observations of a closed-loop DES

The behavior of a DES are *timed events* produced by the combined operation of controller and plant. This behavior is called *internal behavior* of the system. If no information about the internal behavior of a DES is available, the system can be considered as a *black-box*. To identify a model of a black-box system, its external timed behavior needs to be observed. Therefore, the controller input and output signals and the related time is observed by means of the external signals u and t. While the system output u is used to monitor the observable logical behavior of the system, the system time t captures the related time behavior. The observation is implemented such that the DES is *passively* monitored, as shown in Figure 3.2, which means that the observation does not affect the operation of controller and plant. The behavior captured by the external signals represents the *external behavior* of the DES. Due to the event- and state-based nature of these systems, finite state machines are used for modeling in the following.

Definition 5 (Event-driven behavior of the DES). The z-th event $e^{DES}(z)$ produced by the DES at time $t(z)$ causes the system to perform a state transition $(x^{DES}(z-1), x^{DES}(z)) \in R^{DES}$.

The internal behavior of a DES, according to Figure 3.2, is given by events $e^{DES} \in E^{DES}$ and the related time attributes $t \in \mathbb{N}_0^+$. The variable z can be considered as an enumerator to denote the sequence of occurring events $e^{DES}(z)$. When an event $e^{DES}(z)$ occurs, the state of the system is switched to a new state $x^{DES}(z) \in X^{DES}$, with X^{DES}

representing the state-space of the DES. A state x^{DES} represents the combined state of controller and plant. The set of transitions R^{DES} contains pairs of states $(x^{DES}(z-1), x^{DES}(z))$, in which $x^{DES}(z)$ is the successor of $x^{DES}(z-1)$. For a given state $x^{DES}(z-1)$, several outgoing transitions may exist with different succeeding states. In this case, the executed transition is non-deterministically chosen from R^{DES}. The internal time behavior is given by the time attributes t, which represent the time when a state transitions is supposed to occur. The definition of t as a natural number is made for technical reasons and does not restrict the capabilities of the modeling approach. Natural numbers are capable of representing time domains typically used in computation systems. Since these systems will be used for implementation purposes, the natural numbers are an appropriate domain. Events represent actions performed by controller or plant. An action can be, for instance, the movement of a tool, pushing a button, the failure of a hardware component, and the end of a PLC calculation. The effects of events can be recognized by means of the DES output $\Lambda^{DES}(x^{DES}(z))$. An event inducing a state transition is observed if $\Lambda^{DES}(x^{DES}(z)) \neq \Lambda^{DES}(x^{DES}(z-1))$ holds. A state transition from $x^{DES}(z)$ to $x^{DES}(z-1)$ cannot be observed if the output of the closed-loop DES does not change $\Lambda^{DES}(x^{DES}(z)) = \Lambda^{DES}(x^{DES}(z-1))$.

Definition 6 (System output). The output of the DES related to the j-th event is defined as $u(j) = \Lambda^{DES}(x^{DES}(j))$. Two successive outputs $u(j-1)$ and $u(j)$ are distinguishable such that $u(j) \neq u(j-1)$ holds.

The system output represents the output of the DES from an external point of view. In order to distinguish between the internal and external behavior, the new enumerator variable $j = 1, 2, \ldots$ is introduced. The variable j is related to the sequential occurrence of events that can be observed by means of distinguishable system outputs $u(j)$.

Definition 7 (System time). The time t^{DES} when the j-th system output is observed is denoted as the system time $t(j)$.

The system time is *advancing* such that $t(j) > t(j-1)$ holds. This property results from the fact that at a time instance only one event can occur.

Definition 8 (Output time span). The time span of the j-the system output for all $j = 1, 2, \ldots$ is defined as

$$\delta(j) = \begin{cases} t(j) - t(j-1) & \text{if } j > 1 \\ \perp & \text{if } j = 1 \end{cases}. \tag{3.1}$$

The output time span $\delta(j)$ is the time elapsed between the occurrence of a former $u(j-1)$ and a current system output $u(j)$. This time can be interpreted as the *activation time*[4] of the j-th system output $u(j)$. Obviously, $\delta(j)$ is a relative time. For $j = 1$, no preceding system output $u(j-1)$ and no related output time $t(j-1)$ exist. Hence, the output time span is *undefined*, denoted by the bottom type symbol "$\perp$". Due to the advancing time property, $\delta(j) > 0$ for $j > 1$ holds.

[4]In [Cassandras and Lafortune, 2008], $\delta(j)$ is denoted as *lifetime* of the j-th event. Since this term may lead to confusion with the time instance when an event occurs, it will not be used in this thesis.

Definition 9 (Timed system output). The j-th timed system output is defined as

$$u_t(j) = \begin{pmatrix} u(j) \\ \delta(j) \end{pmatrix}. \tag{3.2}$$

The *external behavior* of a closed-loop DES is given as sequences of timed system outputs $u_t(1), u_t(2), u_t(3), \ldots$ that can be observed by means of the external signals u and t.

Internal behavior (black-box)

DES event $e^{DES}(z)$	e_1	e_2	e_3	e_4	e_5	e_6
DES time attribute $t^{DES}(z)$	100	110	120	200	360	410
DES state $x^{DES}(z)$	x_0	x_4	x_5	x_7	x_1	x_2
DES output $\Lambda^{DES}(x^{DES}(z))$	A	A	D	C	C	B
z	1	2	3	4	5	6

External behavior (observable)

system output $u(j)$	A		D	C		B
system time $t(j)$	100		120	200		410
output time span $\delta(j)$	$\perp$		20	80		210
j	1		2	3		4

Figure 3.3: Example of internal and external behavior

Example 3.1 (Internal and external behavior). In Figure 3.3, an example for the relationship between the internal and external behavior of a closed-loop DES is shown. The events e_1, e_3, e_4, and e_6 can be observed by means of the DES output since their occurrence leads the system to enter states with outputs that differ from the ones of the preceding states. The state transitions induced by events e_2 and e_5 do not lead the system output to change, hence the related events are not observable. In general, the number of internal events that can be observed by means of the system output depends on the degree of abstraction and instrumentation of the DES.

Definition 10 (Observed timed output sequences). The set of observed timed output sequences of a DES is given as $\Sigma_t = \{\sigma_{t,1}, \sigma_{t,2}, \ldots, \sigma_{t,h}, \ldots, \sigma_{t,p}\}$ with the h-th sequence given as $\sigma_{t,h} = (u_{t,h}(1), u_{t,h}(2), \ldots, u_{t,h}(l_h))$. A timed system output observed in the h-th sequence is denoted as $u_{t,h}(j)$.

Assumption 1 (Initial timed system output). The initial timed system outputs of all observed timed output sequences are given as $u_{t,i}(1) = u_{t,j}(1)$ for all $\sigma_{t,i}, \sigma_{t,j} \in \Sigma_t$ and for the initial time spans, $\delta_i(1) = \delta_j(1) = \perp$ holds.

The observed timed output sequences represent the data-base that is used for automatic modeling of the DES. Each observed sequence represents a *system evolution*. In

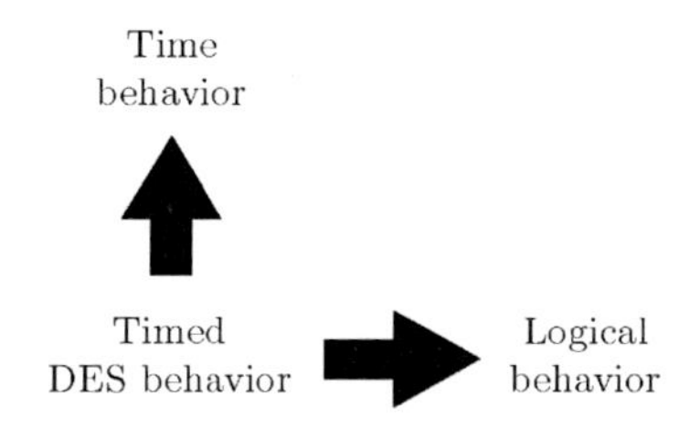

Figure 3.4: Behavior dimensions of a timed DES

production systems, an evolution can represent the execution of a basic production cycle. It is assumed in the following that each observed timed output sequence has the same initial timed system output such that the sequences represent comparable system evolutions. System output $u(j)$ and time spans $\delta(j)$ can be understood as two dimensions of the external timed DES behavior, as shown in Figure 3.4. While the logical behavior specifies the events generated by a DES, the time behavior provides the information when these events occur. In this sense, time behavior can be interpreted as an attribute of the logical behavior.

3.2.2 Timed Model and Languages

In this work, a DES is considered as a timed autonomous event generator. An applied model has to represent the same timed event generating behavior as the original system. Therefore, the *Timed Autonomous Automaton with Output (TAAO)* is introduced in the following. It is a modified Timed Automaton with Guards given in [Cassandras and Lafortune, 2008].

Definition 11 (Timed Autonomous Automaton with Output (TAAO)).

$$TAAO = (X, x_0, c, TG, TT, \Omega, \lambda) \tag{3.3}$$

where X is the finite set of states, x_0 is the initial state, c is the clock with valuation $c(j) \in \mathbb{N}_0^+$, TG is the set of time guards where a time guard is given as the interval $[\tau_{lo}, \tau_{up}]$ with the lower and upper time bounds given as $\tau_{lo}, \tau_{up} \in (\mathbb{N}_0^+ \cup \infty)$, respectively, TT is the set of timed transitions with

$$TT \subseteq X \times TG \times \{c\} \times X$$

where a timed transition is given as $(x, guard, c, x')$ with x' as the successor of state x, $guard \in TG$, the clock c that is to be reset, Ω is the output alphabet and λ: $X \to \Omega$ is the output function.

The notation $x(j)$ is used to denote the model state x that is active after the occurrence of the j-th event. The time related components of the model are given by the clock c and the time guards TG. The clock c runs synchronously to the system time t of the

DES and $c(j)$ represents the relative time since its former reset. By use of the function $guard(x(j-1), x(j))$, the time guard related to transition $(x, guard, c, x')$ is addressed. If $(x(j-1), guard, c, x(j)) \notin TT$, the function $guard(x(j-1), x(j))$ returns the undefined time guard, indicated by $\perp$.

Definition 12 (Model execution rule). A state $x(j)$ is the successor of $x(j-1)$ at time t, if $\exists(x(j-1), guard, c, x(j)) \in TT$ such that $c(j) \in guard(x(j-1), x(j))$ holds. If several possible next state candidates $x(j)$ exist for state $x(j-1)$, given by several possible transitions $TT_{pos} \subseteq TT$ such that $c(j) \in guard(x(j-1), x(j))$, $\forall(x(j-1), guard, c, x(j)) \in TT_{pos}$ holds, the executed transition is chosen non-deterministically from TT_{pos}. After executing a state transition and on model initialization, clock c is reset to zero, i.e. $c(j) := 0$.

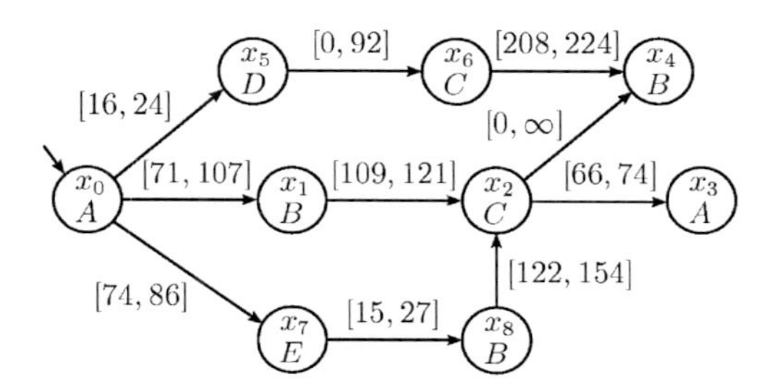

Figure 3.5: Example *TAAO*

In general, the *TAAO* is a non-deterministic automaton with respect to its time and logical behavior. A *TAAO* is *temporally non-deterministic* if at least one transition exists that models a time guard $[\tau_{lo}, \tau_{up}]$ with $\tau_{lo} \neq \tau_{up}$. In this case, the time when the transition fires is non-deterministically selected from $[\tau_{lo}, \tau_{up}]$. If $\tau_{lo} = \tau_{up}$ holds for the time guards $[\tau_{lo}, \tau_{up}]$ of all transitions in TT, then the *TAAO* is temporally deterministic. In this case, there exists for each transition only a single, previously known time instance $\tau = \tau_{lo} = \tau_{up}$ when it is allowed to fire. A *TAAO* is *logically non-deterministic* if at least two transitions exist $(x, guard_1, c, x') \in TT$ and $(x, guard_2, c, x'') \in TT$ that have the same source state x but different successor states x' and x'' and the corresponding time guards $guard_1$ and $guard_2$ are overlapping such that $guard_1 \cap guard_2 \neq \emptyset$ holds. In this case, the executed transition is non-deterministically selected. An example of a temporally and logically non-deterministic *TAAO* is shown in Figure 3.5. It is temporally non-deterministic since for none of the time guards $\tau_{lo} = \tau_{up}$ holds. The logical non-determinism can be recognized, for instance, by the transitions from state x_2 to x_3 and from x_2 to x_4. Both transitions can be executed in the common time interval [66, 74]. In the example *TAAO*, the outputs of the states are represented by capital letters. The applied notation for the state outputs must be equivalent to the notation of system outputs. Another feasible representation, instead of letters, are vectors, for instance. This depends on the application and the nature of the observed system output. Since the clock reset $c(j) := 0$ is performed for all transitions of the $TAAO$, it is omitted in the graphical representation of the automaton.

A $TAAO$ is able to represent the external timed behavior of a DES and can produce the same type of outputs. Observable transitions of the DES $(x^{DES}(j-1), x^{DES}(j)) \in R^{DES}$, which occur at time t^{DES}, are represented by modeled transitions $(x(j-1), guard, c, x(j)) \in TT$ with $guard$ representing the set of output time spans when the transition may occur. The outputs of the DES $u(j) = \Lambda^{DES}(x^{DES}(j))$ are represented by the modeled outputs $u(j) = \lambda(x(j))$. Due to these properties, a $TAAO$ is an appropriate model for timed DES.

In this work, formal languages are used to represent the original, observed, and modeled timed behavior of a DES. A language is a collection of strings while each string is composed by a number of elementary *symbols*. To describe the timed behavior of a DES, the definition of a timed language is introduced in the following. The timed symbols of a closed-loop DES are given by the timed system outputs u_t.

Definition 13 (Timed Word). A timed word w_t^q of length q is defined as

$$w_t^q = \left(\begin{pmatrix} u_1 \\ \delta_1 \end{pmatrix}, \begin{pmatrix} u_2 \\ \delta_2 \end{pmatrix}, \dots, \begin{pmatrix} u_i \\ \delta_i \end{pmatrix}, \dots, \begin{pmatrix} u_q \\ \delta_q \end{pmatrix} \right). \tag{3.4}$$

The related logical word $w^q = (u_1, u_2, \dots, u_i, \dots, u_q)$ is obtained by the *untiming* operation that is defined as

$$w^q = untime(w_t^q). \tag{3.5}$$

The untiming operation 'erases' the time information from the timed string. The notation $w^q[i]$ is used to address the i-th symbol of w^q.

Definition 14 (Original languages). The *original timed language* $L^n_{Orig,t}$ of length n is defined as the set of timed words w_t^q with $1 \leq q \leq n$ that can be generated by the DES. The corresponding *original logical language* L^n_{Orig} is determined by untiming all words of $L^n_{Orig,t}$, according to Equation 3.5.

The original timed language $L^n_{Orig,t}$ is the external, observable, timed behavior of a DES, which the system is able to perform. If the DES is considered as a black-box and no information about it is available, then $L^n_{Orig,t}$ is *unknown*. For timed identification, it is advantageous to distinguish between the logical and the related time behavior. The original logical behavior is modeled by the introduced original logical language L^n_{Orig}. In order to describe the related time behavior, the definition of original time span sets is given in the following.

Definition 15 (Original time span sets). For each original logical word $w^q \in L^n_{Orig}$ of length q with $w^q = (u_1, u_2, \dots, u_i, \dots, u_q)$, there exists a sequence of q *original time span sets*

$$\Delta^{w^q}_{Orig} = (\Delta^{w^q}_{Orig,w^q[1]}, \Delta^{w^q}_{Orig,w^q[2]}, \dots, \Delta^{w^q}_{Orig,w^q[i]}, \dots, \Delta^{w^q}_{Orig,w^q[q]}), \tag{3.6}$$

with $\Delta^{w^q}_{Orig,w^q[i]} \subseteq (\mathbb{N}_0^+ \cup \bot)$, $\forall 1 \leq i \leq q$.

The time span set $\Delta^{w^q}_{Orig,w^q[i]}$ contains the relative times when a symbol u_i of an original logical word $w^q \in L^n_{Orig}$ can be generated by the DES. For a timed word w^q_t, generated by the DES with $w^q = untime(w^q_t) \in L^n_{Orig}$, $\delta_i \in \Delta^{w^q}_{Orig,w^q[i]}$ holds for all symbols u_i of w^q. The time span δ_i, for a given symbol u_i, is an instance of the related original time span set $\Delta^{w^q}_{Orig,w^q[i]}$. The set $\Delta^{w^q}_{Orig,w^q[i]}$ contains all possible time spans δ_i for the symbol u_i of the logical word w^q. Since the original timed language is unknown, $\Delta^{w^q}_{Orig}$ is also unknown $\forall w^q \in L^n_{Orig}$.

Example 3.2 (Original timed language). Given the original timed language

$$L^{n=2}_{Orig,t} = \left\{ \begin{pmatrix} A \\ 5 \end{pmatrix}, \begin{pmatrix} A \\ 7 \end{pmatrix}, \begin{pmatrix} B \\ 1 \end{pmatrix}, \left(\begin{pmatrix} A \\ 5 \end{pmatrix}, \begin{pmatrix} B \\ 1 \end{pmatrix} \right) \right\}$$

which contains words with lengths $1 \leq q \leq 2$. The corresponding original logical language $L^{n=2}_{Orig} = \{A, B, (A, B)\}$ is determined by untiming all words $w^q_t \in L^{n=2}_{Orig,t}$ according to Equation 3.5. The time behavior of $L^{n=2}_{Orig,t}$ is modeled by the original time spans sets, one for each $w^q \in L^{n=2}_{Orig}$. The time spans sets are given as $\Delta^{A}_{Orig} = \{5, 7\}$, $\Delta^{B}_{Orig} = \{1\}$ and $\Delta^{AB}_{Orig} = (\Delta^{AB}_{Orig,A}, \Delta^{AB}_{Orig,B}) = (\{5\}, \{1\})$.

The original logical language L^n_{Orig} and the original time span sets $\Delta^{w^q}_{Orig}$ are an alternative representation of the original timed language $L^n_{Orig,t}$. They are introduced in order to distinguish between the logical and time behavior modeled by a timed language. It has been shown how L^n_{Orig} and $\Delta^{w^q}_{Orig}$ can be determined for a given original timed language $L^n_{Orig,t}$. Since both representations model the same timed behavior, a timed language $L^n_{Orig,t}$ can be unambiguously reconstructed by means of L^n_{Orig} and $\Delta^{w^q}_{Orig}$ for all $w^q \in L^n_{Orig}$. This is done according to the following equation:

$$\begin{aligned} L^n_{Orig,t} = \Bigg\{ w^q_t = \left(\begin{pmatrix} u_1 \\ \delta_1 \end{pmatrix}, \begin{pmatrix} u_2 \\ \delta_2 \end{pmatrix}, \ldots, \begin{pmatrix} u_i \\ \delta_i \end{pmatrix}, \ldots \begin{pmatrix} u_q \\ \delta_q \end{pmatrix} \right) \Bigg| \\ (w^q = (u_1, u_2, \ldots, u_i, \ldots, u_q) \in L^n_{Orig}) \wedge (\delta_i \in \Delta^{w^q}_{Orig,w^q[i]}) \Bigg\} \end{aligned} \quad (3.7)$$

with $\Delta^{w^q}_{Orig,w^q[i]} \in \Delta^{w^q}_{Orig}$. The original timed language $L^n_{Orig,t}$ is the set of timed words w^q_t resulting from the combination of untimed words $w^q \in L^n_{Orig}$ and related time spans from $\Delta^{w^q}_{Orig}$. In Example 3.2, $L^{n=2}_{Orig,t}$ can be reconstructed by combining the logical word A and the time spans from $\Delta^{A}_{Orig} = \{5, 7\}$ to timed words, B and time span of $\Delta^{B}_{Orig} = \{1\}$, and AB and the time spans $\Delta^{AB}_{Orig} = (\{5\}, \{1\})$. The distinction of logical and time behavior will be identically pursued for the observed timed language $L^n_{Obs,t}$ and the modeled timed language $L^n_{Mod,t}$, which are introduced in the following. However, two assumptions are made first concerning the original logical language and the original time span sets.

Assumption 2 (Finite original logical language). The original logical language L^n_{Orig} of the DES is assumed to be finite such that $\exists \varepsilon \in \mathbb{N}$: $|L^n_{Orig}| < \varepsilon$, with $|L^n_{Orig}|$ denoting the number of logical words contained in L^n_{Orig}.

Even if the original logical language of a system has to be considered as unknown, it is reasonable to assume that an (unknown) upper bound ε exists that limits the cardinality

$|L^n_{Orig}|$. This refers to the idea that the amount of logical behavior, a controlled industrial system can perform, is typically limited by constraints such as the physical system setup and the control algorithm. Consequently, an upper bound ε must exist for the number of logical words contained in L^n_{Orig}. As for the original logical language, an assumption is made for the original time span sets.

Assumption 3 (Bounded original time span sets). The original time span sets $\Delta^{w^q}_{Orig} = (\Delta^{w^q}_{Orig,w^q[1]}, \Delta^{w^q}_{Orig,w^q[2]}, \ldots, \Delta^{w^q}_{Orig,w^q[i]}, \ldots, \Delta^{w^q}_{Orig,w^q[q]})$ are assumed to be bounded such that there exists a *lower bound* $\min(\Delta^{w^q}_{Orig,w^q[i]})$ and an *upper bound* $\max(\Delta^{w^q}_{Orig,w^q[i]})$ $\forall 1 \leq i \leq q$ with $\min(\Delta^{w^q}_{Orig,w^q[i]}) \leq \delta_i \leq \max(\Delta^{w^q}_{Orig,w^q[i]})$ $\forall \delta_i \in \Delta^{w^q}_{Orig,w^q[i]}$.

In general, since event timings of a DES are sensitive to noise and disturbances, a large number of output time spans may exist that all belong to the fault-free operation of the system. However, even if there are many different timings, it is assumed that for each original time span set $\Delta^{w^q}_{Orig,w^q[i]}$ a lower and an upper time bound exists. A related symbol u_i is supposed to occur later than the lower bound and earlier than the upper bound. The lower bound refers to the idea that for each event produced by the DES a minimum activation time is required before it can occur. This activation time is typically limited by the physical process such as transportation delays, tool operation times, and calculation times, for instances. The upper bound hast to exist because of the fact that the process continuously makes progress. Every action performed by the DES has to finish after a certain time.

Definition 16 (Observed languages). The set of timed words of length q that is observed after the h-th of p system evolutions is defined as

$$W^q_{Obs,t} = \bigcup_{\sigma_{t,h} \in \Sigma_t} \left(\bigcup_{j=1}^{|\sigma_{t,h}|-q+1} (u_{t,h}(j), u_{t,h}(j+1), \ldots, u_{t,h}(j+q-1)) \right) \tag{3.8}$$

and the *observed timed language* of length n is defined as

$$L^n_{Obs,t} = \bigcup_{q=1}^{n} W^q_{Obs,t}. \tag{3.9}$$

The set of observed logical words W^q_{Obs} and the *observed logical language* L^n_{Obs} are determined by untiming the words of $W^q_{Obs,t}$ and $L^n_{Obs,t}$ according to Equation 3.5, respectively.

Example 3.3 (Observed words). Given the observed timed output sequences $\Sigma_t = \{\sigma_{t,1}, \sigma_{t,2}\}$ as

$$\sigma_{t,1} = \left(\begin{pmatrix} A \\ \bot \end{pmatrix}, \begin{pmatrix} B \\ 80 \end{pmatrix}, \begin{pmatrix} C \\ 112 \end{pmatrix}, \begin{pmatrix} A \\ 72 \end{pmatrix} \right), \sigma_{t,2} = \left(\begin{pmatrix} A \\ \bot \end{pmatrix}, \begin{pmatrix} E \\ 83 \end{pmatrix}, \begin{pmatrix} B \\ 18 \end{pmatrix}, \begin{pmatrix} C \\ 121 \end{pmatrix}, \begin{pmatrix} A \\ 68 \end{pmatrix} \right),$$

the observed timed words of length $q = 1$ are determined as

$$W^{q=1}_{Obs,t} = \left\{ \begin{pmatrix} A \\ \bot \end{pmatrix}, \begin{pmatrix} A \\ 72 \end{pmatrix}, \begin{pmatrix} A \\ 68 \end{pmatrix}, \begin{pmatrix} B \\ 80 \end{pmatrix}, \begin{pmatrix} B \\ 18 \end{pmatrix}, \begin{pmatrix} C \\ 112 \end{pmatrix}, \begin{pmatrix} C \\ 121 \end{pmatrix}, \begin{pmatrix} E \\ 83 \end{pmatrix} \right\}$$

and the corresponding logical words as $W^{q=1}_{Obs} = \{A, B, C, E\}$. The observed timed words of length $q = 2$ result in

$$W^{q=2}_{Obs,t} = \left\{ \left(\begin{pmatrix} A \\ \perp \end{pmatrix}, \begin{pmatrix} B \\ 80 \end{pmatrix} \right), \left(\begin{pmatrix} A \\ \perp \end{pmatrix}, \begin{pmatrix} E \\ 83 \end{pmatrix} \right), \left(\begin{pmatrix} B \\ 80 \end{pmatrix}, \begin{pmatrix} C \\ 112 \end{pmatrix} \right), \left(\begin{pmatrix} B \\ 18 \end{pmatrix}, \begin{pmatrix} C \\ 121 \end{pmatrix} \right), \left(\begin{pmatrix} C \\ 112 \end{pmatrix}, \begin{pmatrix} A \\ 72 \end{pmatrix} \right), \left(\begin{pmatrix} C \\ 121 \end{pmatrix}, \begin{pmatrix} A \\ 68 \end{pmatrix} \right), \left(\begin{pmatrix} E \\ 83 \end{pmatrix}, \begin{pmatrix} B \\ 18 \end{pmatrix} \right) \right\}$$

and the corresponding logical words in $W^{q=2}_{Obs} = \{AB, AE, BC, CA, EB\}$.

Definition 17 (Observed time span sequences). Given an observed word $w^q \in W^q_{Obs}$, the v-long observed time span sequence for the i-th symbol of w^q is given as

$$\Delta^{w^q}_{Obs,w^q[i]} = (\delta_{i,1}, \delta_{i,2}, \ldots, \delta_{i,r}, \ldots, \delta_{i,v}) \tag{3.10}$$

where $\delta_{i,r} = \delta_r(j + i - 1)$ for the r-th observed timed string

$$w^q_r(j) = \left(\begin{pmatrix} u_r(j) \\ \delta_r(j) \end{pmatrix}, \begin{pmatrix} u_r(j+1) \\ \delta_r(j+1) \end{pmatrix}, \ldots, \begin{pmatrix} u_r(j+i-1) \\ \delta_r(j+i-1) \end{pmatrix}, \ldots, \begin{pmatrix} u_r(j+q-1) \\ \delta_r(j+q-1) \end{pmatrix} \right) \tag{3.11}$$

in $\sigma_t \in \Sigma_t$, with $1 \leq i \leq q$ and $1 \leq j \leq |\sigma_t| - q + 1$ such that $untime(w^q_r(j)) = w^q$ holds. The observed time span sequences $\Delta^{w^q}_{Obs}$ for all q symbols of w^q are given as

$$\Delta^{w^q}_{Obs} = (\Delta^{w^q}_{Obs,w^q[1]}, \Delta^{w^q}_{Obs,w^q[2]}, \ldots, \Delta^{w^q}_{Obs,w^q[i]}, \ldots, \Delta^{w^q}_{Obs,w^q[q]}) \tag{3.12}$$

with $\min(\Delta^{w^q}_{Obs,w^q[i]})$ called *lower bound* and $\max(\Delta^{w^q}_{Obs,w^q[i]})$ called *upper bound* of $\Delta^{w^q}_{Obs,w^q[i]}$, respectively. The time span sequences Δ^n_{Obs} for all words $w^q \in L^n_{Obs}$ are given as

$$\Delta^n_{Obs} = \bigcup_{w^q \in L^n_{Obs}} \Delta^{w^q}_{Obs}. \tag{3.13}$$

The operation $|\Delta^{w^q}_{Obs,w^q[i]}|$ determines the number of time spans v contained in the sequence $\Delta^{w^q}_{Obs,w^q[i]}$.

The observed time span sequence $\Delta^{w^q}_{Obs,w^q[i]}$ is defined for each symbol $w^q[i]$ of an observed logical word w^q. In contrast to the original time spans sets, the observed time spans are organized in sequences that allow for duplicated time spans. In order to determine $\Delta^{w^q}_{Obs,w^q[i]}$ for all u_i of a word $w^q \in L^n_{Obs}$, the logical words contained in the observed timed output sequences Σ_t are compared with w^q. If w^q is found within a given $\sigma_t \in \Sigma_t$, the related time spans $\delta(j + i - 1)$ are determined and added to the corresponding time span sequence $\Delta^{w^q}_{Obs,w^q[i]}$. The number of observed time spans v represents the number a word has been observed in all sequences of Σ_t. The value of v may differ from the number of observed timed output sequences p since sequences may exist that do not contain w^q but also sequences may exist in which w^q is contained more than once. The observed timed language $L^n_{Obs,t}$ can be reconstructed out of L^n_{Obs} and Δ^n_{Obs} in the same manner as done by for the original timed language by Equation 3.7.

Example 3.4 (Observed time span sequences). Given the observed timed output sequences $\Sigma_t = \{\sigma_{t,1}, \sigma_{t,2}\}$ and the determined logical words $W_{Obs}^{q=1}$ and $W_{Obs}^{q=2}$ from Example 3.3, the observed time span sequences of word $w^{q=1} = A$ are determined as

$$\Delta_{Obs}^{A} = \Delta_{Obs,A}^{A} = (\perp, 72, \perp, 68)$$

and of $w^{q=2} = CA$ as

$$\Delta_{Obs}^{CA} = (\Delta_{Obs,C}^{CA}, \Delta_{Obs,A}^{CA}) = ((112, 121), (72, 68))$$

with $\Delta_{Obs,C}^{CA}$ representing the timings of the symbol C and $\Delta_{Obs,A}^{CA}$ of A that occurred in the context of $w^{q=2} = CA$.

The timings of a symbol depend on the considered word. A symbol u_i contained in different words w_1^q and w_2^q may refer to different time span sequences $\Delta_{Orig,i}^{w_1^q} \neq \Delta_{Orig,i}^{w_2^q}$. In Example 3.4, the time span sequences $\Delta_{Obs,A}^{A}$ and $\Delta_{Obs,A}^{CA}$ represent both time spans of symbol A, however, they differ from each other since $\Delta_{Obs,A}^{CA}$ considers only those timings of A in case that A occurred right after C.

Assumption 4 (Convergence of L_{Obs}^n to L_{Orig}^n). With increasing number of observed system evolutions p, it is assumed that L_{Obs}^n converges to L_{Orig}^n.

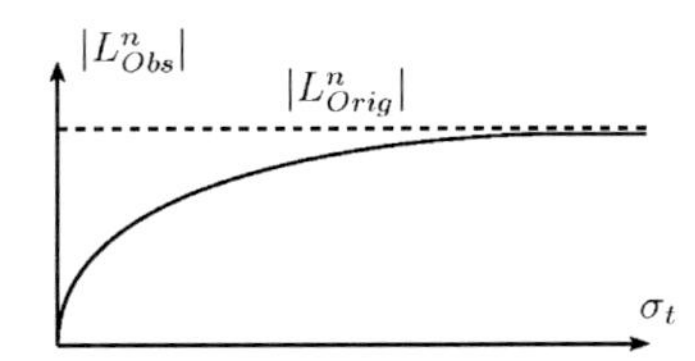

Figure 3.6: Assumed convergence of the observed logical language L_{Obs}^n

In Figure 3.6, the convergence of the observed to the unknown, original logical language of a DES is depicted. The upper bound of $|L_{Orig}^n|$, according to Assumption 2, is illustrated by a dotted line. With increasing number of observed timed output sequences $\sigma_t \in \Sigma_t$, it is assumed that $|L_{Obs}^n|$ converges against $|L_{Orig}^n|$. If $|L_{Obs}^n|$ remains stable for a significant number of observations, then $|L_{Obs}^n|$ is assumed to be converged and the original logical language L_{Orig}^n of the DES is considered as completely observed. The number of output sequences that need to be observed until $|L_{Obs}^n|$ is converged depends on the behavior of the DES and can be very large, for instance in case of concurrency.

Assumption 5 (Convergence of the observed to the original time span bounds). Given words of the observed logical language $w^q \in L_{Obs}^n$ and the corresponding observed time span sequences $\Delta_{Obs}^{w^q} = (\Delta_{Obs,w^q[1]}^{w^q}, \Delta_{Obs,w^q[2]}^{w^q}, \dots, \Delta_{Obs,w^q[q]}^{w^q})$, with increasing number of observed system evolutions p, it is assumed that

1. $\min(\Delta_{Obs,w^q[i]}^{w^q})$ converges to $\min(\Delta_{Orig,w^q[i]}^{w^q})$ and

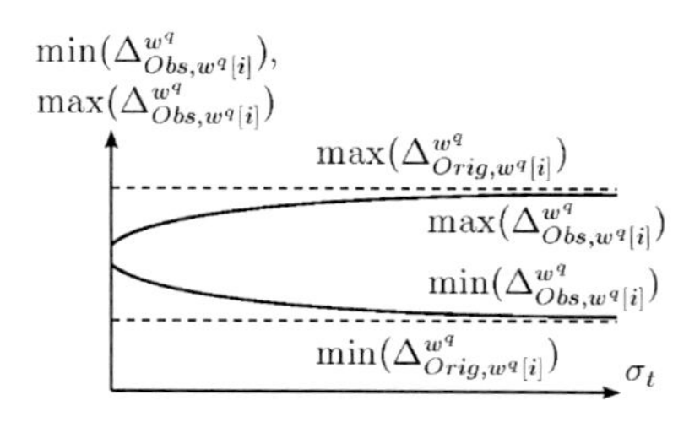

Figure 3.7: Assumed convergence of bounds for the i-th observed time span sequence $\Delta^{w^q}_{Obs,w^q[i]}$ related to the logical word w^q

2. $\max(\Delta^{w^q}_{Obs,w^q[i]})$ converges to $\max(\Delta^{w^q}_{Orig,w^q[i]})$

$\forall 1 \leq i \leq q$ with $\bot \notin \Delta^{w^q}_{Obs,w^q[i]}$.

In contrast to the convergence of the logic sequential language, which is defined with respect to its cardinality, the convergence of the time spans is defined with respect to the bounds. This principle is depicted in Figure 3.7. Due to the large number of possible original timings that exist in general, it is not reasonable to expect that they can all be observed within an acceptable observation horizon. Instead, $\Delta^{w^q}_{Obs,w^q[i]}$ contains *samples* of the original time span sets $\Delta^{w^q}_{Orig,w^q[i]}$. It is assumed that with growing number of observations, the minimum and maximum bounds of the observed time span sequences $\Delta^{w^q}_{Obs,w^q[i]}$ converge against the bounds of the corresponding original time spans sets. In accordance with this, it is possible to define an interval based on the observed bounds in which all original time spans are supposed to be found. This holds for observed time span sequences that do not contain undefined time spans.

After the definition of the original and observed languages and the associated time span sets and sequences, the definitions of the modeled languages and the modeled time span sets follow. These definitions represent the modeled logic and time behavior a *TAAO* can reproduce.

Definition 18 (Modeled language). Given a *TAAO*, the modeled timed words of length q for a given state $x(j)$ are defined as

$$W^{q=1,x(j)}_{Mod,t} = \left\{ w^{q=1}_t \middle| \left(w^{q=1}_t = \begin{pmatrix} \lambda(x(j)) \\ \delta(j) \end{pmatrix} \right) \wedge (\delta(j) \in guard(x(j-1), x(j))) \right\}, \quad (3.14)$$

$$\begin{aligned} W^{q>1,x(j)}_{Mod,t} = \Bigg\{ w^q_t \Bigg| w^q_t = & \left(\begin{pmatrix} \lambda(x(j)) \\ \delta(j) \end{pmatrix}, \begin{pmatrix} \lambda(x(j+1)) \\ \delta(j+1) \end{pmatrix}, \ldots, \begin{pmatrix} \lambda(x(j+q-1)) \\ \delta(j+q-1) \end{pmatrix} \right) \wedge \\ & (\delta(j) \in guard(x(j-1), x(j))) \wedge \\ & (\delta(j+1) \in guard(x(j), x(j+1))) \wedge \ldots \wedge \\ & (\delta(j+q-1) \in guard(x(j+q-2), x(j+q-1))) \wedge \\ & (x(i-1), guard, c, x(i)) \in TT \; \forall j+1 \leq i \leq j+q-1 \Bigg\}. \end{aligned} \quad (3.15)$$

The modeled timed words of length q for all states result in

$$W^q_{Mod,t} = \bigcup_{x \in X} W^{q,x}_{Mod,t} \tag{3.16}$$

and the *modeled timed language* of length n is determined as

$$L^n_{Mod,t} = \bigcup_{q=1}^{n} W^q_{Mod,t}. \tag{3.17}$$

The *modeled logical language* L^n_{Mod} is determined by untiming all words of $L^n_{Mod,t}$ according to Equation 3.5.

Example 3.5. Consider the example automaton given in Figure 3.5. The set of modeled timed words $W^{q=1}_{Mod,t}$ of length $q = 1$ is determined as

$$W^{q=1}_{Mod,t} = \left\{ \begin{pmatrix} A \\ \perp \end{pmatrix}, \begin{pmatrix} A \\ 66 \end{pmatrix}, \dots, \begin{pmatrix} A \\ 74 \end{pmatrix}, \begin{pmatrix} B \\ 0 \end{pmatrix}, \dots, \begin{pmatrix} B \\ \infty \end{pmatrix}, \begin{pmatrix} C \\ 0 \end{pmatrix}, \dots, \begin{pmatrix} C \\ 92 \end{pmatrix}, \begin{pmatrix} C \\ 109 \end{pmatrix}, \dots, \begin{pmatrix} C \\ 154 \end{pmatrix}, \begin{pmatrix} D \\ 16 \end{pmatrix}, \dots, \begin{pmatrix} D \\ 24 \end{pmatrix}, \begin{pmatrix} E \\ 74 \end{pmatrix}, \dots, \begin{pmatrix} E \\ 86 \end{pmatrix} \right\}$$

and the set of logic sequential words as $W^{q=1}_{Mod} = \{A, B, C, D, E\}$. The set of modeled timed words $W^{q=2}_{Mod,t}$ of length $q = 2$ is determined as

$$W^{q=2}_{Mod,t} = \left\{ \left(\begin{pmatrix} A \\ \perp \end{pmatrix} \begin{pmatrix} B \\ 71 \end{pmatrix} \right), \dots, \left(\begin{pmatrix} A \\ \perp \end{pmatrix} \begin{pmatrix} B \\ 107 \end{pmatrix} \right), \left(\begin{pmatrix} B \\ 71 \end{pmatrix} \begin{pmatrix} C \\ 109 \end{pmatrix} \right), \dots, \left(\begin{pmatrix} B \\ 107 \end{pmatrix} \begin{pmatrix} C \\ 121 \end{pmatrix} \right), \left(\begin{pmatrix} C \\ 109 \end{pmatrix} \begin{pmatrix} A \\ 66 \end{pmatrix} \right), \dots \right\}$$

and the set of logic sequential words as $W^{q=2}_{Mod} = \{AD, DC, CB, AB, BC, CA, AE, EB\}$.

The timed words modeled by a *TAAO* are given by the outputs of states $\lambda(x(j))$ and the time spans $\delta(j) \in guard(x(j-1), x(j))$ related to transitions that end in the corresponding states $x(j)$. In that sense, a time span $\delta(j)$ is the *activation time* of the j-th model output. While a set of modeled logical words W^q_{Mod} contains always a finite number of logical words, the number of timed words modeled by $W^q_{Mod,t}$ can be infinite. Therefore, at least one transition has to exist with $\tau_{up} = \infty$.

Definition 19 (Modeled time span sets). Given a modeled logical word w^q with $w^q = (u_1, u_2, \dots, u_i, \dots, u_q) \in W^q_{Mod}$, the modeled time span set $\Delta^{w^q}_{Mod,w^q[i]}$ of the i-th symbol of w^q is defined as

$$\Delta^{w^{q=1}}_{Mod} = \{\delta \mid (\delta \in guard(x(j-1), x(j))) \wedge (\lambda(x(j) = u_1)\}, \tag{3.18}$$

$$\begin{aligned} \Delta^{w^{q>1}}_{Mod,w^q[i]} = \{ & \delta \mid \delta \in guard(x(j+i-2), x(j+i-1)) \wedge \\ & (x(r-1), guard, c, x(r)) \in TT \; \forall j+1 \leq r \leq j+q-1 \wedge \\ & (\lambda(x(j)) = u_1) \wedge (\lambda(x(j+1)) = u_2) \wedge \dots \wedge (\lambda(x(j+q-1)) = u_q)\} \end{aligned} \tag{3.19}$$

with $1 \leq i \leq q$ and for all q symbols as

$$\Delta_{Mod}^{w^{q>1}} = (\Delta_{Mod,w^q[1]}^{w^{q>1}}, \Delta_{Mod,w^q[2]}^{w^{q>1}}, \ldots, \Delta_{Mod,w^q[i]}^{w^{q>1}}, \ldots, \Delta_{Mod,w^q[q]}^{w^{q>1}}). \quad (3.20)$$

The set of modeled time span for all words $w^q \in L_{Mod}^n$ is given as

$$\Delta_{Mod}^{n} = \bigcup_{w^q \in L_{Mod}^n} \Delta_{Mod}^{w^q}. \quad (3.21)$$

The modeled time language $L_{Mod,t}^n$ can be reconstructed out of L_{Mod}^n and Δ_{Mod}^n according to Equation 3.7, as it has been presented for the original timed language.

Example 3.6 (Modeled time span sets). Consider again the example automaton given in Figure 3.5. The modeled time span set of the modeled logical word $w^{q=1} = C$ results in

$$\Delta_{Mod}^{C} = \{[0, 92]\} \cup \{[109, 121]\} \cup \{[122, 154]\}.$$

The time spans sets are given by the time guards that are associated to transitions ending in states with $\lambda(x) = C$. The modeled time spans of the word $w^{q=2} = CA$ can be determined as

$$\Delta_{Mod}^{CA} = \left(\Delta_{Mod,C}^{CA}, \Delta_{Mod,A}^{CA}\right) = \left(\{[109, 121]\} \cup \{[122, 154]\}, \{[66, 74]\}\right).$$

3.3 Identification of Timed Models

3.3.1 Time Identification Approach

The aim of timed identification is to automatically determine a *TAAO* based on observed timed output sequences Σ_t of the DES. In order to identify a model of the fault-free system behavior, the following assumption needs to be made:

Assumption 6 (Fault-free observed timed output sequences). The observed timed output sequences Σ_t are samples of the *fault-free* original DES behavior.

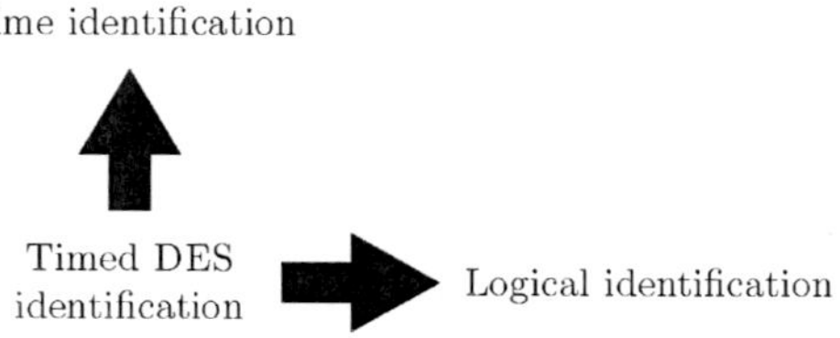

Time identification

Timed DES identification

Logical identification

Figure 3.8: Dimensions of the timed DES identification

The identification approach follows the timed behavior definitions presented in the preceding section. Based on the observed logical language L_{Obs}^n and the related observed time span sequences Δ^n_{Obs}, the timed language of the model $L^n_{Mod,t}$ is identified. The identification approach proposed in the following performs both *logical* and *time identification*. While the modeled logical language L^n_{Mod} is identified based on L^n_{Obs}, the modeled time span sets Δ^n_{Mod} result from time identification based on Δ^n_{Obs}. This two-dimensional identification principle is illustrated in Figure 3.8. The procedure to identify the logical behavior of the model relies on an approved algorithm from literature, originally proposed in [Klein, 2005] and improved in [Roth, 2010]. The presented algorithm is based on an identification parameter $k \geq 1$. An automation is automatically built that represents the original logical language L^n_{Orig} of the DES. To identify a model that additionally represents the original time behavior, the algorithm is extended in this thesis by appropriate time identification procedures. Therefore, a new approach for time identification is introduced in the following.

Time identification is the automatic determination of all time guards $guard(x, x')$ of the *TAAO* based on the observed time spans Δ^n_{Obs}. Time guard determination is performed for a logical identified automaton in which states and logical transitions are already defined. A logical identified transition is denoted as $(x, -, -, x')$ with the source state x and the target state x'. The time guard and the clock are not defined yet, indicated by the don't care symbol "$-$". The procedures for time identification does not affect the modeled logical language L^n_{Mod} of a *TAAO* but only the modeled time behavior Δ^n_{Mod}.

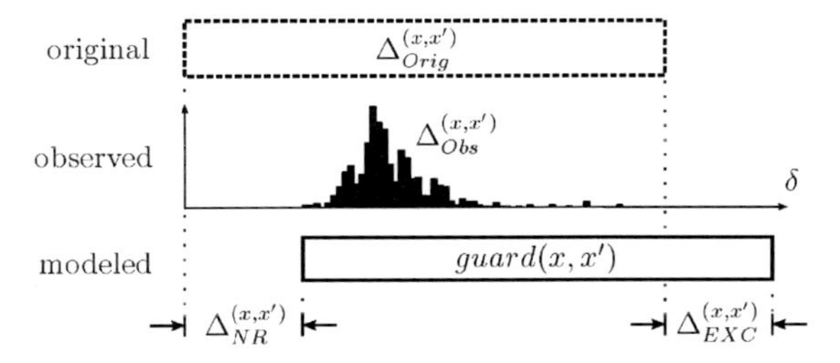

Figure 3.9: Original, observed and modeled transition time spans

For each logically identified transition $(x, -, -, x') \in TT$, a time identification problem, depicted in Figure 3.9, can be formulated. Given a sequence of *observed transition time spans* $\Delta^{(x,x')}_{Obs}$, the problem is to determine a time guard $guard(x, x')$ such that the corresponding set of *original transitions time spans* $\Delta^{(x,x')}_{Orig}$ is properly represented. Before a time guard can be identified, the observed transition time spans are determined as $\Delta^{(x,x')}_{Obs} = \Delta^{w^q}_{Obs,w^q[i]}$ with $\Delta^{w^q}_{Obs,w^q[i]}$ denoting the observed time span sequence related to transition $(x, -, -, x')$. This task, called *time span allocation*, is part of the timed identification algorithm and will be explained when introducing the algorithm later on. For the following explanation, it is assumed that the time span allocation is already done and the observed transition time spans $\Delta^{(x,x')}_{Obs}$ are available for all logically identified transitions. After allocation, time guard determination is performed with respect to the *exceeding* $\Delta^{(x,x')}_{EXC}$ and *non-reproducible transition time spans* $\Delta^{(x,x')}_{NR}$. These time span sets have to be minimized since they lead to the undesired exceeding and non-reproducible timed languages, introduced in Chapter 2. The set of exceeding time spans of a transition from state x to x' is defined as $\Delta_{EXC}^{(x,x')} = \{guard(x,x')\} \setminus \Delta_{Orig}^{(x,x')}$. The set $\Delta_{EXC}^{(x,x')}$ represents time spans modeled by the time guard that cannot be generated by the DES. The set of non-reproducible time spans is given by $\Delta_{NR}^{(x,x')} = \Delta_{Orig}^{(x,x')} \setminus \{guard(x,x')\}$. These time spans represent original time spans that cannot be reproduced by the model. Since $\Delta_{Orig}^{(x,x')}$ is a priori unknown, $\Delta_{EXC}^{(x,x')}$ and $\Delta_{NR}^{(x,x')}$ cannot be explicitly determined. In literature, a commonly used approach is to determine the lower τ_{lo} and upper bound τ_{up} of $guard(x,x')$ by the minimum and maximum observed bounds of $\Delta_{Obs}^{(x,x')}$, independent of the number of available observations, see [Supavatanakul et al., 2006] for instance. In that way, $\Delta_{EXC}^{(x,x')}$ is minimized. However, this procedure may lead to a large number of non-reproducible time spans $\Delta_{NR}^{(x,x')}$ if the bounds of the observed transition time span set are not converged to the original ones. In this case, a possibility to reduce $\Delta_{NR}^{(x,x')}$ is to enlarge the observation horizon. Since this is not always possible, the following approach called *tolerance extension* is introduced. The approach allows to extend time guards according to the number of observations v and a tuning parameter ext_0 such that they can be completely identified, especially in the case that the bounds of the observed transition time span sets are not converged yet.

Due to the influence of noise and disturbances, time spans cannot be predicted and appear randomly from an external point of view. Consequently, the observed transition time spans $\Delta_{Obs}^{(x,x')}$ are considered in the following as the result of a *random experiment*. In Figure 3.9, $\Delta_{Obs}^{(x,x')}$ is depicted as a frequency distribution. In various literature approaches it is assumed that original transition time spans $\Delta_{Orig}^{(x,x')}$ follow given probability distributions, e.g. normal distributions in [Das and Holloway, 2000] and normal and exponential distributions in [Lefebvre and Leclercq, 2011]. However, since for the considered DES no knowledge about the distribution of $\Delta_{Orig}^{(x,x')}$ is available, such assumptions cannot be justified in general. Instead, non-parametric approaches for time guard determination have to be preferred that do not rely on a distribution assumption. In the following, the *non-parametric tolerance interval* according to [Hartung and Klösener, 2009] is introduced. It is given as

$$\left[\min\left(\Delta_{Obs}^{(x,x')}\right), \max\left(\Delta_{Obs}^{(x,x')}\right)\right] \tag{3.22}$$

with $\min(\Delta_{Obs}^{(x,x')})$ representing the smallest and $\max(\Delta_{Obs}^{(x,x')})$ representing the largest observed time span of transition $(x,-,-,x')$. The non-parametric tolerance interval is defined with respect to a confidence $1-\alpha$ and a minimum hit rate β. In such an interval, at least $\beta \cdot (1-\alpha)$ of the future observed time spans will be found. In order to determine a tolerance interval for $1-\alpha$ and β, a minimum number of time spans v_0 need to be observed. This number is determined, according to [Hartung and Klösener, 2009], by the following equation.

$$v_0 = \frac{1}{4}\chi^2_{4;1-\alpha} \cdot \frac{1+\beta}{1-\beta} + \frac{1}{2} \tag{3.23}$$

with $\chi^2_{4;1-\alpha}$ the $1-\alpha$-quantile of the chi-squared distribution with 4 degrees of freedom. If

$v \geq v_0$, the non-parametric tolerance interval is a proper estimate of the original transition time spans with respect to $1-\alpha$ and β.[5] However, if a tolerance interval is determined based on less observations $v < v_0$, it has to be expected that the original transition time spans are not completely represented, leading to non-reproducible time spans $\Delta_{NR}^{(x,x')}$. In order to encounter this problem, the tolerance interval is extended beyond the minimum and maximum observations. The size of the extension depend on

1. the number of available observations v and
2. the width of the observed frequency distribution $\left(\max(\Delta_{Obs}^{(x,x')}) - \min(\Delta_{Obs}^{(x,x')})\right)$.

The first rule respects the number of available time spans for the determination of the tolerance interval. The more observations are available, the less the tolerance interval needs to be extended. This stays in accordance with Assumption 5, which states that the observed bounds converge against the original bounds with increasing number of observations. The second rule adapts the extension according to the spread of a distribution. Wide spreading distributions require for more tolerance than distributions that have only little spread. Based on the presented rules, the equation to determine the *tolerance extension* is given as

$$ext\left(\Delta_{Obs}^{(x,x')}\right) = \begin{cases} \infty & \text{if } v = 1 \\ ext_0 \cdot \left(1 - \frac{v}{v_o}\right) \cdot \left(\max\left(\Delta_{Obs}^{(x,x')}\right) - \min\left(\Delta_{Obs}^{(x,x')}\right)\right) & \text{if } 1 < v < v_0 \\ 0 & \text{if } v \geq v_0 \end{cases} \tag{3.24}$$

with the selectable extension parameter ext_0. The tolerance extension is linear to the number of available observations v and to the spread of the time span distribution estimated by $\max(\Delta_{Obs}^{(x,x')}) - \min(\Delta_{Obs}^{(x,x')})$. If $v = 1$, it is not reasonable to define an interval since no information about the spread can be determined by means of the observation. Hence, $ext\left(\Delta_{Obs}^{(x,x')}\right) = \infty$ such that the corresponding transition is not constrained with respect to time. In the case that $v \geq v_0$ or $ext_0 = 0$, then $ext\left(\Delta_{Obs}^{(x,x')}\right) = 0$.

3.3.2 Timed Identification Algorithm

In the following, the timed identification algorithm that identifies the timed behavior of a *TAAO* is presented. The algorithm combines the procedures for logical identification by [Roth, 2010] and the time identification approach proposed in this thesis. For a clear understanding of the new algorithm, all relevant logical identification steps from [Roth, 2010] are recalled and the time identification is explained in detail. Initially, the observed timed output sequences have to be modified. The first symbol of each sequences are duplicated $k-1$-times according to the following equation given in [Roth, 2010]:

$$\sigma_{t,h}^k(i) = \begin{cases} \sigma_{t,h}(1) & \text{for } 1 \leq i \leq k \\ \sigma_{t,h}(i-k+1) & \text{for } k < i \leq (k + |\sigma_{t,h}| - 1) \end{cases} \tag{3.25}$$

[5] Recall that v represents the number of observed transition time spans $v = |\Delta_{Obs}^{(x,x')}|$.

The resulting duplicated sequences are denoted as $\Sigma_t^k = \{\sigma_{t,1}^k, \sigma_{t,2}^k, \ldots, \sigma_{t,h}^k, \ldots, \sigma_{t,p}^k\}$ in the following.

Example 3.7. Given the observed timed sequence

$$\sigma_t = \left(\begin{pmatrix} A \\ \perp \end{pmatrix}, \begin{pmatrix} B \\ 80 \end{pmatrix}, \begin{pmatrix} C \\ 112 \end{pmatrix}, \begin{pmatrix} A \\ 72 \end{pmatrix} \right)$$

and the logical identification parameter $k = 2$, the $k-1$-times duplicated sequence is determined as

$$\sigma_t^{k=2} = \left(\begin{pmatrix} A \\ \perp \end{pmatrix}, \begin{pmatrix} A \\ \perp \end{pmatrix}, \begin{pmatrix} B \\ 80 \end{pmatrix}, \begin{pmatrix} C \\ 112 \end{pmatrix}, \begin{pmatrix} A \\ 72 \end{pmatrix} \right).$$

Algorithm 1 Timed identification algorithm

Require: Observed timed output sequences Σ_t, logical identification parameter k, time identification parameters $1-\alpha, \beta, ext_0$

1: Build W_{Obs}^{k,Σ_t^k} and W_{Obs}^{k+1,Σ_t^k} using Equations 3.25, 3.8, and 3.5
2: Build $\Delta_{Obs}^{k+1,\Sigma_t^k}$ for W_{Obs}^{k+1,Σ_t^k} using Equation 3.13
3: Identify the logical structure and allocate the observed time spans by Algorithm 2
4: Determine the time guards by Algorithm 3
5: **return** $TAAO$

The timed identification is performed according to Algorithm 1. Given the observed timed output sequences and the identification parameters, the first step is to build the observed logic sequential words W_{Obs}^{k,Σ_t^k}, W_{Obs}^{k+1,Σ_t^k}, and the related time span sets $\Delta_{Obs}^{q,\Sigma_t^k}$ based on the duplicated sequences Σ_t^k.[6] In Line 3, Algorithm 2 is executed to logically identify the states, transitions, and the outputs of states and to allocate the observed time spans to the corresponding transitions. After determining the time guards in Line 4 of Algorithm 3, the identified $TAAO$ is returned.

For the following explanations, it is advantageous to introduce a function that determines a substring of a logical word for given positions a and b. The substring function is defined as

$$w^q\langle a..b\rangle = (u(a), u(a+1), \ldots, u(b)), \tag{3.26}$$

with $w^q = (u(1), u(2), \ldots, u(q)) \in W^q$, $a < b \leq q$ and $a, b \in \mathbb{N}$.

Example 3.8 (Substring function). Given the word $w^{q=4} = ABCA$, the substring for $a = 2$ and $b = 3$ is given as $w^{q=4}\langle 2..3\rangle = BC$

The logical identification, the allocation of time spans, and the output determination is performed by Algorithm 2. The observed logical words W_{Obs}^{k,Σ_t^k} are used to build the states of the automaton. For each logical word $w^k \in W_{Obs}^{k,\Sigma_t^k}$, a new state is created in

[6]The index Σ_t^k, contained in the notation of word and time span sets, indicates in the following that these sets are built upon the duplicated sequences.

Algorithm 2 Logical identification and time span allocation

Require: Observed logical words W_{Obs}^{k,Σ_t^k}, W_{Obs}^{k+1,Σ_t^k}, observed time spans $\Delta_{Obs}^{k+1,\Sigma_t^k}$, logical identification parameter k

1: $X := \emptyset$, $x_0 := \emptyset$, $TT := \emptyset$, $\Omega := \emptyset$
2: **for all** $w^k \in W_{Obs}^{k,\Sigma_t^k}$ **do**
3: Create new state x with $\lambda(x) = w^k$
4: $X := X \cup x$
5: **end for**
6: **for all** $w^{k+1} \in W_{Obs}^{k+1,\Sigma_t^k}$ **do**
7: Create transition $(x, -, -, x')$ with $(\lambda(x) = w^{k+1}\langle 1..k\rangle) \wedge (\lambda(x') = w^{k+1}\langle 2..k+1\rangle)$
8: Allocate the transition time spans $\Delta_{Obs}^{(x,x')} := \Delta_{Obs,w^{k+1}[k+1]}^{w^{k+1},\Sigma_t^k}$ for $\Delta_{Obs}^{w^{k+1},\Sigma_t^k} = (\Delta_{Obs,w^{k+1}[1]}^{w^{k+1},\Sigma_t^k}, \Delta_{Obs,w^{k+1}[2]}^{w^{k+1},\Sigma_t^k}, \ldots, \Delta_{Obs,w^{k+1}[k+1]}^{w^{k+1},\Sigma_t^k}) \in \Delta_{Obs}^{k+1,\Sigma_t^k}$
9: $TT := TT \cup (x, -, -, x')$
10: **end for**
11: Set initial state $x_0 := x$ with $(x \in X)$: $(\lambda(x)[i] = untime(\sigma_{t,1}(1))\ \forall 1 \leq i \leq k)$
12: **for all** $x \in X$ **do**
13: Determine the output $\lambda(x) = \lambda(x)[k]$
14: $\Omega := \Omega \cup \lambda(x)$
15: **end for**

Line 3, and the output of the state is set to the string given by the considered word. In that way, each word $w^k \in W_{Obs}^{k,\Sigma_t^k}$ is represented by an unique state. The transitions are generated based on the logical words W_{Obs}^{k+1,Σ_t^k} in Line 7. A logical identified transition $(x, -, -, x')$ connects two states x and x' such that the output of x is equal to the first k-long substring of w^{k+1} and the output of x' is equal to the latter k-long substring of w^{k+1}. The k-long substrings correspond to the outputs of the modeled states that have been identified based on the words of length k. Since a single state can be unambiguously assigned to each word of length k, the two states that are connected by a transition can be unambiguously determined. After creating the logical part of the transition, the related time spans are allocated in Line 8. For each logical word $w^{k+1} \in W_{Obs}^{k+1,\Sigma_t^k}$, which has been used to identify the transitions, a collection of time spans $\Delta_{Obs}^{w^{k+1},\Sigma_t^k} \in \Delta_{Obs}^{k+1,\Sigma_t^k}$ exists. The relevant timing information of the corresponding transition is contained in the $(k+1)$-th element of $\Delta_{Obs}^{w^{k+1},\Sigma_t^k}$ given by $\Delta_{Obs,w^{k+1}[k+1]}^{w^{k+1},\Sigma_t^k}$. In Line 11, the initial state of the *TAAO* is determined by selecting the state x whose output represents the k-times duplication of the initial symbol of the first observed sequence. The algorithm ends with the final output determination. In Line 13, the output of each state is set to the last symbol of the represented word. All other symbols have no impact on the output of the model and can therefore be masked out.

The tolerance extension approach for the identification of the time guards is implemented by Algorithm 3. Initially, the required number of observations v_0 for valid tolerance intervals is determined based on the time identification parameters $1-\alpha$ and β.

Algorithm 3 Time guard determination

Require: Observed transition time spans $\Delta_{Obs}^{(x,x')}$, time identification parameters $1-\alpha, \beta, ext_0$

1: $TG := \emptyset$
2: Determine v_0 by Equation 3.23
3: **for all** $(x, -, -, x') \in TT$ **do**
4: $\quad \tau_{lo} := \min\left(\Delta_{Obs}^{(x,x')}\right) - ext\left(\Delta_{Obs}^{(x,x')}\right)$
5: $\quad \tau_{lo} := \max\left(0, \lfloor \tau_{lo} \rfloor\right)$
6: $\quad \tau_{up} := \max\left(\Delta_{Obs}^{(x,x')}\right) + ext\left(\Delta_{Obs}^{(x,x')}\right)$
7: $\quad \tau_{up} := \lceil \tau_{up} \rceil$
8: $\quad$ Append time guard $TG := TG \cup [\tau_l, \tau_u]$
9: $\quad$ Redefine transition $(x, -, -, x') := (x, [\tau_{lo}, \tau_{up}], c, x')$
10: **end for**

Then, the time guards for each transition are derived. The lower time bound τ_{lo} of a time guard is determined as the minimum observed time span reduced by the tolerance extension $ext(\Delta_{Obs}^{(x,x')})$, which is determined according to Equation 3.24. The final lower time bound τ_{lo} is given by the maximum value of 0 and the determined lower time bound that is rounded down. This avoids negative and non-integer time bounds. The upper time bound τ_{up} is calculated respectively by adding the tolerance interval extension to the maximum observed time span value. The determined value is rounded up in Line 7. Finally, the transitions are redefined with respect to the new time guards that are given by the adapted non-parametric tolerance intervals.

Example 3.9 (Timed identification). The following example illustrates the timed identification of a $TAAO$ according to the algorithms proposed in this work. The observed timed output sequences are given as $\Sigma_t = \{\sigma_{t,1}, \sigma_{t,2}, \sigma_{t,3}, \sigma_{t,4}, \sigma_{t,5}, \sigma_{t,6}\}$ with

$$\sigma_{t,1} = \left(\begin{pmatrix} A \\ \perp \end{pmatrix}, \begin{pmatrix} B \\ 80 \end{pmatrix}, \begin{pmatrix} C \\ 112 \end{pmatrix}, \begin{pmatrix} A \\ 72 \end{pmatrix}\right), \sigma_{t,2} = \left(\begin{pmatrix} A \\ \perp \end{pmatrix}, \begin{pmatrix} B \\ 98 \end{pmatrix}, \begin{pmatrix} C \\ 118 \end{pmatrix}, \begin{pmatrix} B \\ 76 \end{pmatrix}\right),$$
$$\sigma_{t,3} = \left(\begin{pmatrix} A \\ \perp \end{pmatrix}, \begin{pmatrix} D \\ 18 \end{pmatrix}, \begin{pmatrix} C \\ 8 \end{pmatrix}, \begin{pmatrix} B \\ 212 \end{pmatrix}\right), \sigma_{t,4} = \left(\begin{pmatrix} A \\ \perp \end{pmatrix}, \begin{pmatrix} D \\ 22 \end{pmatrix}, \begin{pmatrix} C \\ 64 \end{pmatrix}, \begin{pmatrix} B \\ 220 \end{pmatrix}\right),$$
$$\sigma_{t,5} = \left(\begin{pmatrix} A \\ \perp \end{pmatrix}, \begin{pmatrix} E \\ 77 \end{pmatrix}, \begin{pmatrix} B \\ 24 \end{pmatrix}, \begin{pmatrix} C \\ 146 \end{pmatrix}\right), \sigma_{t,6} = \left(\begin{pmatrix} A \\ \perp \end{pmatrix}, \begin{pmatrix} E \\ 83 \end{pmatrix}, \begin{pmatrix} B \\ 18 \end{pmatrix}, \begin{pmatrix} C \\ 130 \end{pmatrix}, \begin{pmatrix} A \\ 68 \end{pmatrix}\right).$$

and the logical identification parameter as $k = 2$. The $(k-1)$-times duplicated sequences Σ_t^k are determined as:

$$\sigma_{t,1}^{k=2} = \left(\begin{pmatrix}A\\\perp\end{pmatrix}, \begin{pmatrix}A\\\perp\end{pmatrix}, \begin{pmatrix}B\\80\end{pmatrix}, \begin{pmatrix}C\\112\end{pmatrix}, \begin{pmatrix}A\\72\end{pmatrix}\right),$$
$$\sigma_{t,2}^{k=2} = \left(\begin{pmatrix}A\\\perp\end{pmatrix}, \begin{pmatrix}A\\\perp\end{pmatrix}, \begin{pmatrix}B\\98\end{pmatrix}, \begin{pmatrix}C\\118\end{pmatrix}, \begin{pmatrix}B\\76\end{pmatrix}\right),$$
$$\sigma_{t,3}^{k=2} = \left(\begin{pmatrix}A\\\perp\end{pmatrix}, \begin{pmatrix}A\\\perp\end{pmatrix}, \begin{pmatrix}D\\18\end{pmatrix}, \begin{pmatrix}C\\8\end{pmatrix}, \begin{pmatrix}B\\212\end{pmatrix}\right),$$
$$\sigma_{t,4}^{k=2} = \left(\begin{pmatrix}A\\\perp\end{pmatrix}, \begin{pmatrix}A\\\perp\end{pmatrix}, \begin{pmatrix}D\\22\end{pmatrix}, \begin{pmatrix}C\\64\end{pmatrix}, \begin{pmatrix}B\\220\end{pmatrix}\right),$$
$$\sigma_{t,5}^{k=2} = \left(\begin{pmatrix}A\\\perp\end{pmatrix}, \begin{pmatrix}A\\\perp\end{pmatrix}, \begin{pmatrix}E\\77\end{pmatrix}, \begin{pmatrix}B\\24\end{pmatrix}, \begin{pmatrix}C\\146\end{pmatrix}\right),$$
$$\sigma_{t,6}^{k=2} = \left(\begin{pmatrix}A\\\perp\end{pmatrix}, \begin{pmatrix}A\\\perp\end{pmatrix}, \begin{pmatrix}E\\83\end{pmatrix}, \begin{pmatrix}B\\18\end{pmatrix}, \begin{pmatrix}C\\130\end{pmatrix}, \begin{pmatrix}A\\68\end{pmatrix}\right).$$

In the next step, the duplicated observed logical words W_{Obs}^{k,Σ_t^k} and W_{Obs}^{k+1,Σ_t^k} and the observed time span sequences $\Delta_{Obs}^{k+1,\Sigma_t^k}$ have to be derived. They result in

$$\begin{aligned}
W_{Obs}^{k=2,\Sigma^2} &= \{AA, AB, BC, CA, CB, AD, DC, AE, EB\},\\
W_{Obs}^{k+1=3,\Sigma^2} &= \{AAB, ABC, BCA, BCB, AAD, ADC, DCB, AAE, AEB, EBC\},\\
\Delta_{Obs}^{k+1=3,\Sigma^2} &= \{\Delta_{Obs}^{AAB,\Sigma^2}, \Delta_{Obs}^{ABC,\Sigma^2}, \Delta_{Obs}^{BCA,\Sigma^2}, \Delta_{Obs}^{BCB,\Sigma^2}, \Delta_{Obs}^{AAD,\Sigma^2}, \Delta_{Obs}^{DCB,\Sigma^2}, \Delta_{Obs}^{AAE,\Sigma^2},\\
&\quad \Delta_{Obs}^{ADC,\Sigma^2}, \Delta_{Obs}^{AEB,\Sigma^2}, \Delta_{Obs}^{EBC,\Sigma^2}\}
\end{aligned}$$

with

$$\begin{aligned}
\Delta_{Obs}^{AAB,\Sigma^2} &= ((\perp,\perp),(\perp,\perp),(80,98)), & \Delta_{Obs}^{ABC,\Sigma^2} &= ((\perp,\perp),(80,98),(112,118)),\\
\Delta_{Obs}^{BCA,\Sigma^2} &= ((80,18),(112,130),(72,68)), & \Delta_{Obs}^{BCB,\Sigma^2} &= ((98),(118),(76)),\\
\Delta_{Obs}^{AAD,\Sigma^2} &= ((\perp,\perp),(\perp,\perp),(18,22)), & \Delta_{Obs}^{ADC,\Sigma^2} &= ((\perp,\perp),(18,22),(8,64)),\\
\Delta_{Obs}^{DCB,\Sigma^2} &= ((18,22),(8,64),(212,220)), & \Delta_{Obs}^{AAE,\Sigma^2} &= ((\perp),(\perp),(77,83)),\\
\Delta_{Obs}^{AEB,\Sigma^2} &= ((\perp),(77,83),(24,18)), & \Delta_{Obs}^{EBC,\Sigma^2} &= ((77,83),(24,18),(146,130)).
\end{aligned}$$

Based on the sets of words $W_{Obs}^{k=2,\Sigma^2}$, $W_{Obs}^{k+1=3,\Sigma^2}$, the sets of time spans contained in $\Delta_{Obs}^{k+1=3,\Sigma^2}$, and the logical identification parameter $k = 2$, Algorithm 2 is applied. Figure 3.10 shows the identified automaton after executing Line 11 of the Algorithm. The final outputs of states, determined in the remaining steps of the algorithm, are given by the last letters of the words that belong to these states. These letters are marked bold in the figure. The transition time spans are given by the sequences that are allocated to each transition. The size of these sequences can differ from transition to transition according to the number of made observations.

The time guards are identified using Algorithm 3. Given the allocated transition time spans $\Delta_{Obs}^{(x,x')}$ and the time identification parameters $1-\alpha = 0.99$, $\beta = 0.99$, and $ext_0 = 0.5$,

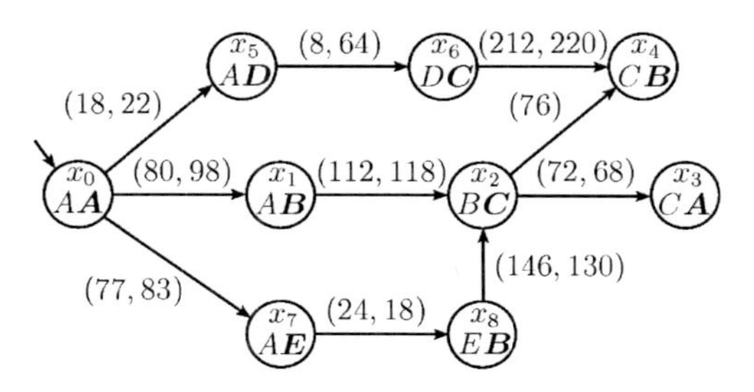

Figure 3.10: Example *TAAO* after logical identification and time span allocation

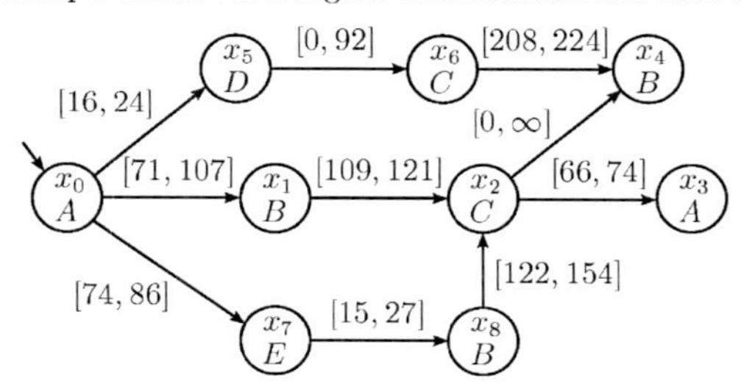

Figure 3.11: Timed identified example *TAAO*

the *TAAO* shown is Figure 3.11 is finally obtained. The number of required observations is determined as $v_0 = 8775$ with $\chi^2_{4;0.99} = 13.28$. $\chi^2_{4;0.99}$ can be found in tables in literature, e.g. in [Hartung and Klösener, 2009]. Since in this example none of the observed time span collections contain enough observations and $ext_0 > 0$, all tolerance intervals are extended. Obviously, the *TAAO* in Figure 3.11 is able to reproduce all observed timed output sequences in Σ_t. The observed and modeled logical words of length $q = 4$ are determined as:

$$W^{q=4}_{Obs} = \{ABCA, ABCB, ADCB, AEBC, EBCA\},$$
$$W^{q=4}_{Mod} = \{ABCA, ABCB, ADCB, AEBC, EBCA, \boldsymbol{EBCB}\}.$$

This shows that the model is able to reproduce all observed logical words of length $q = 4$. However, $EBCB$ represents a modeled logical word that has not been observed. This word belongs either to the original logical behavior of the DES or represents exceeding logical behavior. The observed and modeled time spans for word $w^{q=3} = EBC$ can be determined as:

$$\Delta^{EBC}_{Obs} = ((77, 83), (24, 18), (146, 130)),$$
$$\Delta^{EBC}_{Mod} = ([74, 86], [15, 27], [122, 154]).$$

One can see that the observed time spans are contained in the modeled time span intervals, respectively. The observed time behavior can be reproduced by the model. However, each modeled time span set $\Delta^{EBC}_{Mod,EBC[i]}$ contains timings that are beyond $\min(\Delta^{EBC}_{Obs,EBC[i]})$ and $\max(\Delta^{EBC}_{Obs,EBC[i]})$. Again, depending on the DES, these timings refer either to the original time behavior or represent exceeding time behavior.

The timed identification algorithm has polynomial computation complexity. The calculation efforts to determine $W^{k,\Sigma^k_t}_{Obs}$, $W^{k+1,\Sigma^k_t}_{Obs}$, and $\Delta^{k+1,\Sigma^k_t}_{Obs}$ depend mainly on the number of observed timed output sequences Σ_t and on the length l_h of each observed sequence $\sigma_{t,h} \in \Sigma_t$. The remaining calculations are mainly performed within the for-loops of the Algorithms 2 and 3. The number of iterations results in $2 \cdot |W^{k,\Sigma^k_t}_{Obs}| + 2 \cdot |W^{k+1,\Sigma^k_t}_{Obs}|$ for state and output determination and for transition and time guard identification, respectively.

In the presented time identification algorithm, the tolerance intervals are extended symmetrically with respect to $\min(\Delta^{(x,x')}_{Obs})$ and $\max(\Delta^{(x,x')}_{Obs})$. Depending on the application, time span distributions may exist that reveal a significant degree of *skewness* such that symmetric extension of tolerance intervals may be disadvantageous due to poor fitting. For this case, a modified time guard determination approach is proposed in [Schneider et al., 2012], which considers skewness in particular by asymmetric adaption of the tolerance extensions.

3.3.3 Precision and Completeness Properties

In order to discuss precision properties of the identified automata *TAAO*, *logical* and *time precision* are considered separately in the following. The logical precision of the identified *TAAO* refers to a theorem that has been proofed in [Roth, 2010]. An automaton identified with the presented logic identification algorithm is logically precise in the sense that $\forall n \leq k+1$, $L^n_{Mod} = L^n_{Obs}$. This means that for a given logical identification parameter k, the model exactly reproduces the observed logical words for a length lower or equal to $k+1$. Such an identified *TAAO* has no exceeding logical language with respect to words of length $n \leq k+1$. For words of length $n > k+1$, this cannot be guaranteed. Logically precise models are preferably used for model-based fault diagnosis since the number of missed logical faults is minimized.

The time precision of an identified *TAAO* is defined with respect to the observed and modeled time bounds. It has been shown that the observed timed span sequences of a DES typically represent an incomplete collection of samples of the original time behavior. This results from the randomness that affects the generation of the original time behavior. Therefore, time precision is defined with respect to the observed and modeled time bounds instead of considering single samples, as it is done for the logical precision. An identified *TAAO* according to the proposed timed identification algorithm is temporally precise according to the following theorem.

Theorem 1. If a *TAAO* is logically precise and timed identified such that $|\Delta^{(x,x')}_{Obs}| \geq v_0$ $\forall (x, guard, c, x') \in TT$ or $ext_0 = 0$, then $\forall w^q \in L^{k+1}_{Mod}$, the q-th identified time span set $\Delta^{w^q}_{Mod,w^q[q]}$ of $\Delta^{w^q}_{Mod} = (\Delta^{w^q}_{Mod,w^q[1]}, \Delta^{w^q}_{Mod,w^q[2]}, \ldots, \Delta^{w^q}_{Mod,w^q[q]})$ is *precisely bounded*, such that $\min\left(\Delta^{w^q}_{Mod,w^q[q]}\right) = \min\left(\Delta^{w^q}_{Obs,w^q[q]}\right)$ and $\max\left(\Delta^{w^q}_{Mod,w^q[q]}\right) = \max\left(\Delta^{w^q}_{Obs,w^q[q]}\right)$ holds.

The proof of the theorem is given in the Appendix A. In words, given a logically precise automaton, if a sufficient number of time spans are observed for each transition or the time identification parameter is selected as $ext_0 = 0$, then no extension is added to the tolerance interval. In this case, the bounds of the q-th modeled time span set $\Delta^{w^q}_{Mod,w^q[q]}$ for word w^q *exactly* represent the observed bounds of $\Delta^{w^q}_{Obs,w^q[q]}$. The modeled time span sets contain no timings that are smaller than $\min\left(\Delta^{w^q}_{Obs,w^q[q]}\right)$ and larger than $\max\left(\Delta^{w^q}_{Obs,w^q[q]}\right)$. Logical and time precision helps to minimize the exceeding timed language $L^n_{EXC,t}$ of a model and to reduce the number of missed faults. It has to be remarked that a necessary condition for Theorem 1 is that tolerance intervals are not extended during identification. Any extension induced by $ext_0 > 0$ for transitions with $v < v_0$ decreases the time given precision of an identified *TAAO*.

The *completeness* of an identified *TAAO* is analyzed for the combined logical and timed behavior. The following theorem shows that a *TAAO*, identified according to the presented timed identification algorithm, can simulate the original timed language of a DES.

Theorem 2. If $L^{k+1}_{Orig} = L^{k+1}_{Obs}$ and $\forall w^q \in L^{k+1}_{Obs}$, $\min\left(\Delta^{w^q}_{Orig,w^q[i]}\right) = \min\left(\Delta^{w^q}_{Obs,w^q[i]}\right)$ and $\max\left(\Delta^{w^q}_{Orig,w^q[i]}\right) = \max\left(\Delta^{w^q}_{Obs,w^q[i]}\right)$ $\forall 1 \leq i \leq q$ with $\perp \notin \Delta^{w^q}_{Orig,w^q[i]}$, then $L^n_{Mod,t} \supseteq L^n_{Orig,t}$ with $n \geq 1$ for an identified *TAAO*.

The proof of the theorem is given in the Appendix A. In words, if the original logical language up to length $k+1$ and the original time span bounds for all symbols of all observed logical words are completely observed, then the modeled time language $L^n_{Mod,t}$ simulates the original timed language $L^n_{Orig,t}$ for any word with length $n \geq 1$. The first necessary condition $L^{k+1}_{Orig} = L^{k+1}_{Obs}$ refers to the logical completeness of an identified *TAAO*. It has been proven in [Roth, 2010] that this property holds for automata that are identified based on the presented logical identification procedures. In order to identify an automaton that can reproduce logical words of an arbitrary length, the logical original language of the system with length $k+1$ must be completely observed. This is important, because it allows to identify a model by means of an observed language with comparable shorter length in order to reproduce original words of comparable greater length. Note that logical languages with small lengths require in general fewer observations to converge than logical languages with greater length. The time completeness is ensured by the second condition, which requires the complete observation of the original time span bounds for all observed words. In particular, if v_0 observations are made, the time bounds can be considered as converged with respect to $1-\alpha$ and β. If a sufficient number of observations is not available, the tolerance extension approach allows to artificially enlarge the tolerance intervals such that the number of non-reproducible time spans is reduced. By ensuring the logical and time completeness, the timed language $L^n_{Mod,t}$, composed of modeled logical words of length up to n, and the corresponding modeled time spans simulate the original timed language $L^n_{Orig,t}$ of the DES. Logical and time completeness help to minimize the non-reproducible timed language L^n_{NR} that can lead to false detections during fault diagnosis.

One can see that precision and completeness are in general contrary requirements. Improving one property by an appropriate selection of the identification parameters typically leads to degrade the other one. The *trade-off* between model precision and completeness has to be made by the identification engineer with respect to the model application. Therefore, the meaning and impact of the identification parameters are discussed in the following.

3.3.4 Identification Parameters

Logic identification parameter k

The logical identification of a *TAAO* is based on the logical identification parameter k. The parameter defines the length of words W_{Obs}^{k} and W_{Obs}^{k+1} that are used to determine states, outputs, and logically identified transitions of an automaton. Depending on the requirements for model precision and completeness, a lower and an upper bound for k can be determined. The lower bound for k refers to the minimal logical precision that is required for an identified automaton. Since $L_{Mod}^{n} = L_{Obs}^{n}$ holds for words with $n \leq k+1$, increasing k decreases the number of words contained in the exceeding language. However, k cannot be arbitrarily increased in general since an upper bound for the logical identification parameter k exists which is related to logical completeness. The condition for logical completeness of a *TAAO* is given as $L_{Orig}^{k+1} = L_{Obs}^{k+1}$. The parameter k needs to be chosen such that the original language of length $k+1$ can be assumed as completely observed. Since L_{Orig}^{k+1} is unknown, the completeness of L_{Obs}^{k+1} is estimated by means of the convergence of $|L_{Obs}^{k+1}|$. According to Assumption 4, the observed logical language converges to the original logical language with growing number of observations. Given k, if $|L_{Obs}^{k+1}|$ has converged after a number of observed timed output sequences to a stable level, then L_{Orig}^{k+1} can be assumed to be completely observed. In this case, it is most likely that no new words of length $n \leq k+1$ will be discovered in future observations. The upper bound for k is determined by the language L_{Obs}^{k+1} which can be considered as completely observed while languages for larger values of k can no longer be considered as completely observed for the given sequence of observations. The identification engineer can reveal this bound by analyzing the growths of the observed languages. In summary, selecting a small k allows to identify models with minor non-reproducible logical behavior while selecting a large k reduce the exceeding logical behavior. For practical application it is suggested to select the logical identification parameter k as the maximum value under the condition that $|L_{Obs}^{k+1}|$ can be considered as converged for the given sequence of observations. Typical values for k are 1, 2, and 3. A comprehensive study of the logical identification parameter is given in [Roth, 2010].

Time identification parameters $1-\alpha$, β, and ext_0

The time identification is performed with respect to the parameters $1-\alpha$, β, and ext_0. The statistical parameters $1-\alpha$ and β come along with the definition of the non-parametric tolerance interval. In particular, for a given transition and with the probability of $1-\alpha$,

β of all future observed time spans that belong to the fault-free behavior of a system are contained in the tolerance interval. Since the tolerance intervals constitute the time guards that are supposed to represent the entire future observed fault-free time behavior, the probabilities $1-\alpha$ and β should be selected very high. In that way, real fault-free behavior observed in future is reproduced by the time guards with high probability. Typical values for $1-\alpha$ and β are 0.95, 0.975 and 0.99. This selection is made independent of the considered application. However, since the parameters define the representative properties of the tolerance interval, they are actually no tuning parameters of the identification algorithm. In contrast to that, the interval extension parameter ext_0 represents the 'real' tuning parameter. Depending on the application, ext_0 is selected to balance precision and completeness of time guards when the required number of observations v_0 is not available. In order to determine an upper bound for ext_0 the worst case scenario needs to be considered. This scenario is given with the time identification of a transition, with $v=2$ such that the width of the time guard $w_{tg} = \tau_{up} - \tau_{lo}$ results in $w_{tg} = (1 + 2 \cdot ext_0) \cdot w_{tol}$ and the width of the tolerance interval is given as $w_{tol} = \max\left(\Delta_{Obs}^{(x,x')}\right) - \min\left(\Delta_{Obs}^{(x,x')}\right)$. For instance, if $ext_0 = 0.5$, then $w_{tg} = 2 \cdot w_{tol}$. In the worst case, for $v = 2$ and choosing $ext_0 = 0.5$, the resulting time guard is twice the width of the determined tolerance interval. An identification engineer needs to decide, depending on the application, whether this imprecision due to the additional tolerance can be accepted in the worst case. The lower bound of ext_0 refers to the amount of non-reproducible behavior that can be tolerated. This value can be determined by validating the model based on additional observed data, which is different to the data used for time identification. Since the model is used for FDI in the following, a validation can be performed with respect to the number of false detections that result from the non-reproducible behavior. In practice, the tuning parameter can be initially chosen as $ext_0 = 0$. After validation, ext_0 can then be successively increased until an acceptable value for the non-reducible behavior is reached. Note that the time guards of transitions with $v \geq v_0$ are not affected by the selection of ext_0. A comprehensive validation of the time identification parameters by means of the BMS is given in Chapter 5.

3.4 Timed Distributed Modeling

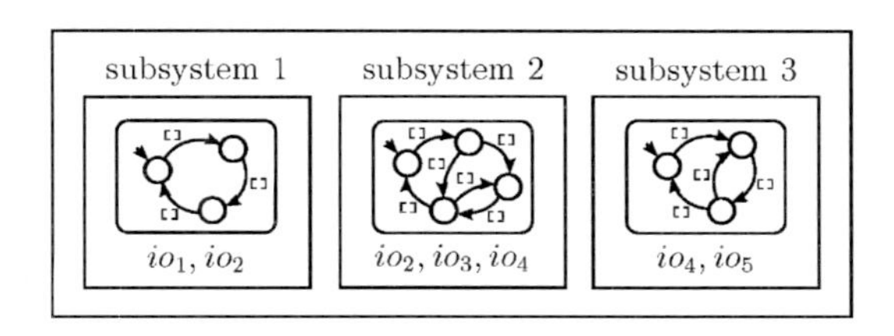

Figure 3.12: Structure of a DES with concurrent operating subsystems

Many real-world DES are *concurrent* systems that are composed of a number of concurrently operating *subsystems* [Willner and Heymann, 1991]. In Figure 3.12, an example of a concurrent DES is depicted. A subsystem represents a part of the entire DES, for instance, a manufacturing tool, a conveyor, or a single sensor. Each subsystem can be characterized by the sensors and actuators that are attached to the considered part of the system. By means of the controller I/Os, which are connected to the related sensors and actuators of the plant, it is possible to clearly distinguish between the subsystems and to describe their partial output behavior. In the figure, the I/Os related to subsystem 1 are io_1 and io_2, for instance. The partial behavior generated by each subsystem can be described by means of a partial timed automaton. The global behavior of a DES is given by the combined behavior of all subsystems.

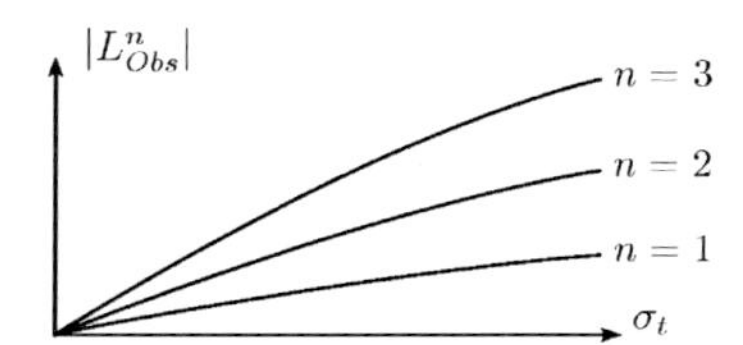

Figure 3.13: Observed language growth of an exemplary concurrent DES

The timed modeling and identification approach, presented so far, focused on the identification of timed *monolithic* models. In order to identify a complete timed monolithic model of a DES, the logical completeness must be ensured. It is assumed that a logically complete model can be identified if the cardinality of the observed logical language, which is used for identification, is converged to a stable level. In this case, the original behavior of the DES is, most likely, completely observed and no new observations are expected in future. As discussed in the preceding section, it needs to be investigated before identification whether this convergence holds for the applied observed languages. In Figure 3.13, the observed language growth of an exemplary concurrent DES is shown. Obviously, none of the observed logical languages, for any lengths n, can be considered as converged for the given set of observed output sequences. Hence, an identified monolithic model, based on this data set, will be likely incomplete for any choice of the logical identification parameter k. This is characteristic for concurrent systems. Since the combined behavior of all subsystems needs to be observed for the complete observation, a given set of observed sequences may not be sufficient. Another problem, which arises in the context of monolithic identification of concurrent DES, is the number of states that are necessary for modeling the combined behavior. This refers to the well-known state-space explosion problem that often comes along with modeling of large systems.

One way to explicitly consider concurrency during modeling is to use appropriate models, for instance Petri nets, which are able to represent concurrent systems by nature. This is addressed in the works of [Maruster et al., 2003] and [Medeiros et al., 2005], for instance. Another possibility is to use a modeling structure in which each concurrent operating subsystem is modeled by an individual *partial model*. In that way, the behavior of the DES is represented by the combined behavior of all partial models. It allows to use models for subsystems that do not represent concurrent behavior in an explicit manner, e.g. automata. This modeling concept is pursued in [Philippot et al., 2007] in the field of DES modeling for fault diagnosis purposes. In this thesis, a *timed distributed modeling* framework is applied, in which the subsystems of the DES are modeled by partial timed automata. In that way, the internal concurrency of each partial model is minimized. Distributed modeling is especially advantageous for concurrent system that need to be completely modeled for a given, likely incomplete set of observed timed output sequences. By using an appropriate distributed model, the observed logical languages of the partial models converge for a given set of observed timed output sequences while the observed logical language related to the monolithic model may not. In order to identify a timed distributed model of a closed-loop DES, the distributed structure of the model needs to be defined. This is done in the following by means of the *I/O-partition*. Before the definition is introduced, the set of controller inputs and outputs is formalized.

Definition 20 (Controller I/Os). The set of controller I/Os of a closed-loop DES is defined as $IO = \{io_1, io_2, \ldots, io_m\}$, each $io_i \in IO$ representing either a controller input or controller output and m denotes the number of all I/Os.

Definition 21 (I/O-partition). Given a closed-loop DES with m controller inputs and outputs $IO = \{io_1, io_2, \ldots, io_m\}$, the I/O-partition is defined as

$$P = \{SUB_1, SUB_2, \ldots, SUB_N\} \tag{3.27}$$

with the i-th I/O-subset $SUB_i \subseteq IO$ and $|SUB_i|$ denotes the number of represented I/Os. An I/O-partition P is *valid*, i.e.

$$IO = \bigcup_{SUB_i \in P} SUB_i \tag{3.28}$$

holds.

An I/O-subset SUB_i delimits a subsystem by representing the related I/Os. In contrast to the mathematical definition of a partition, in which all I/O-subsets are mutually exclusive, Definition 21 allows for *shared I/Os*. An $io \in IO$ is a shared I/O, if $\exists SUB_i \in P$ and $\exists SUB_j \in P$, $i \neq j$, such that $io \in SUB_i$ and $io \in SUB_j$. Shared I/Os synchronize two or more subsystems with respect to their generated behavior. Two I/O-subsets SUB_i and SUB_j, $i \neq j$, are called *overlapping* if they share at least one I/O, i.e. $SUB_i \cap SUB_j \neq \emptyset$. A monolithic model is based on the trivial I/O-partition $P = \{IO\}$. An I/O-partition is called valid if the I/O-subsets SUB of P cover all I/Os IO of the DES. In the following, valid I/O-partitions are considered, exclusively.

Distributed modeling requires a redefinition of the system output and the system time. According to Definition 6, the external logical behavior of a DES is observed by means of the system output u. In the following, the system output u represents the *controller I/O-vector* of the DES. For a closed-loop DES, with m controller inputs and outputs

$IO = \{io_1, io_2, \ldots, io_m\}$, the controller I/O-vector u is given as

$$u = \begin{pmatrix} io_1 \\ io_2 \\ \vdots \\ io_m \end{pmatrix}. \tag{3.29}$$

If one or more I/Os change their value, then a new system output $u(j)$ is generated and each element of the vector represents the current value of an I/O. The behavior of a subsystem, however, is defined with respect to a subset of all I/Os. Consequently, a partial system output must be defined for each subsystem such that only the corresponding I/Os, given by the I/O-subset, are considered.

Definition 22 (Partial system output). Given the set of controller inputs and outputs $IO = \{io_1, io_2, \ldots, io_m\}$, the partial system output related to the I/O-subset $SUB \subseteq IO$ is defined as

$$u_{SUB} = \begin{pmatrix} u_{SUB}[1] \\ u_{SUB}[2] \\ \vdots \\ u_{SUB}[m] \end{pmatrix} \tag{3.30}$$

with

$$u_{SUB}[i] = \begin{cases} u[i] & \text{if } io_i \in SUB \\ - & \text{if } io_i \notin SUB \end{cases}. \tag{3.31}$$

The operator "$-$" is used as don't care symbol. The j-th partial system output is denoted as $u_{SUB}(j)$.

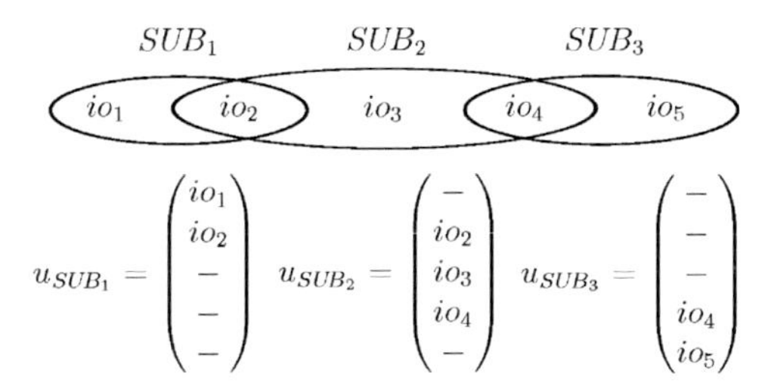

Figure 3.14: Example I/O-partition and partial system outputs

Two successive partial system outputs are distinguishable such that $u_{SUB}(j) \neq u_{SUB}(j-1)$ holds. In Figure 3.14, an example of partial system outputs for a given I/O-partition is shown. One can see that each vector contains only those I/Os that belong to the related I/O-partition subset. The vector elements of the remaining I/Os are replaced by don't care symbols. In the monolithic case, with $SUB = IO$, the partial system output turns into the system output for the entire closed-loop DES. The partial system time $t_{SUB}(j)$ refers to the occurrence of the j-th partial system output. $t_{SUB}(j)$ is advancing $t_{SUB}(j) > t_{SUB}(j-1)$, just as in the monolithic case.

Definition 23 (Partial timed system output).

$$u_{SUB,t}(j) = \begin{pmatrix} u_{SUB}(j) \\ \delta_{SUB}(j) \end{pmatrix} \tag{3.32}$$

$\delta_{SUB}(j)$ is the relative time passed since the occurrence of the preceding partial system output $u_{SUB}(j-1)$, in accordance with Definition 8.

In order to determine the observed timed output sequences of a subsystem $\Sigma_{SUB,t}$, by means of the observed timed output sequences Σ_t of the DES, the timed projection function is defined in the following:

Definition 24 (Timed projection function). Given an I/O-subset $SUB \subseteq IO$ and an observed timed output sequence $\sigma_t \in \Sigma_t$, the timed projection function $proj_{SUB,t}(\sigma_t)$ replaces each $u_t(j)$ in σ_t by the corresponding partial timed system output $u_{SUB,t}(j)$ and subsequently, all partial timed system outputs $u_{SUB,t}(j)$ with $u_{SUB}(j) = u_{SUB}(j-1)$ are removed and $\delta_{SUB}(j+1) = \delta_{SUB}(j+1) + \delta_{SUB}(j)$. The projected output sequence is denoted as *partial timed output sequence* $\sigma_{SUB,t}$.

The timed projection operation 'erases' timed system outputs in which the subsystem SUB_i is not involved and adapts the resulting relative time spans, accordingly. In that way, the determined partial timed output sequences $\Sigma_{SUB,t}$ contain distinguishable successive partial outputs and the resulting time spans remain consistent to the global time behavior.

Example 3.10 (Timed projection function). Given the controller inputs and outputs $IO = \{io_1, io_2, io_3\}$, the I/O-subset $SUB = \{io_1, io_2\}$, and the timed output sequence

$$\sigma_t = \left(\begin{pmatrix} \begin{pmatrix} 0 \\ 0 \\ 0 \end{pmatrix} \\ \bot \end{pmatrix}, \begin{pmatrix} \begin{pmatrix} 0 \\ 0 \\ 1 \end{pmatrix} \\ 10 \end{pmatrix}, \begin{pmatrix} \begin{pmatrix} 1 \\ 1 \\ 1 \end{pmatrix} \\ 25 \end{pmatrix}, \begin{pmatrix} \begin{pmatrix} 1 \\ 0 \\ 0 \end{pmatrix} \\ 17 \end{pmatrix} \right)$$

the partial timed output sequence $\sigma_{SUB,t}$ is determined as

$$proj_{SUB,t}(\sigma_t) = \left(\begin{pmatrix} \begin{pmatrix} 0 \\ 0 \\ - \end{pmatrix} \\ \bot \end{pmatrix}, \begin{pmatrix} \begin{pmatrix} 1 \\ 1 \\ - \end{pmatrix} \\ 35 \end{pmatrix}, \begin{pmatrix} \begin{pmatrix} 1 \\ 0 \\ - \end{pmatrix} \\ 17 \end{pmatrix} \right).$$

In the first step of the projection, the timed system outputs $u_t(j)$ in σ_t are replaced by the partial timed system outputs $u_{SUB,t}(j)$. Since $io_3 \notin SUB$, the third element of $u_{SUB}(j)$ is set as "$-$". Then, the resulting partial timed system outputs $u_{SUB,t}(j)$ are revised. Since $u_{SUB}(1) = u_{SUB}(2)$ holds after the replacement,

$$u_{SUB,t}(2) = \begin{pmatrix} \begin{pmatrix} 0 \\ 0 \\ - \end{pmatrix} \\ 10 \end{pmatrix}$$

is removed and $\delta_{SUB}(3) = 25 + 10 = 35$. The relative time between $u_{SUB}(1)$ and the next partial system output $u_{SUB}(3)$ is updated to $\delta_{SUB}(3) = 35$. Since the remaining partial system outputs are different from the preceding one, respectively, no further removing and time span updating is done. Due to the removal of $u_{SUB,t}(2)$, the resulting partial timed output sequence given by $proj_{SUB,t}(\sigma_t)$ contains one vector less than the original given output sequences σ_t.

Definition 25 (Timed distributed model). Given a DES and an I/O-partition $P = \{SUB_1, SUB_2, \ldots, SUB_N\}$, the *timed distributed model* is defined as

$$TDM = \{TAAO_1, TAAO_2, \ldots, TAAO_N\} \tag{3.33}$$

with $TAAO_i$ as the *partial* $TAAO$ related to the I/O-subset $SUB_i \in P$ and the partial timed system output $u_{TAAO_i,t}$.[7]

The timed distributed model is based on the I/O-partition P. Each $SUB_i \in P$ is represented by a $TAAO$ according to Definition 11. The combined behavior of all partial automata models the global timed behavior of the considered DES. The original and observed partial languages are defined according Definitions 14 and 16 with respect to the partial timed output vector $u_{SUB_i,t}$ and to the projected partial timed output sequences $\Sigma_{SUB_i,t}$. Likewise, the original partial time span sets and observed partial time span sequences are defined according to Definitions 15 and 17. In that way, the partial models can be identified with the approaches presented for a monolithic $TAAO$.

3.5 Identification of Timed Distributed Models

3.5.1 Timed Distributed Identification Approach

The timed distributed identification approach is visualized in Figure 3.15. Given an I/O-partition P and the observed timed output sequences Σ_t, the observed partial timed output sequences $\Sigma_{SUB_i,t}$ are determined for each $SUB_i \in P$ according to Definition 24. The projected timed output sequences represent the observed timed behavior of the DES with respect to the considered subsystem, given by SUB_i. They are the data-base for the identification of the partial timed automata. Given the identification parameters, the timed identification according to Algorithm 1 is executed for each I/O-subsets SUB_i

[7] The notations $u_{SUB_i,t}$ and $u_{TAAO_i,t}$ represent both the same partial timed system output. They are synonymously used in the following.

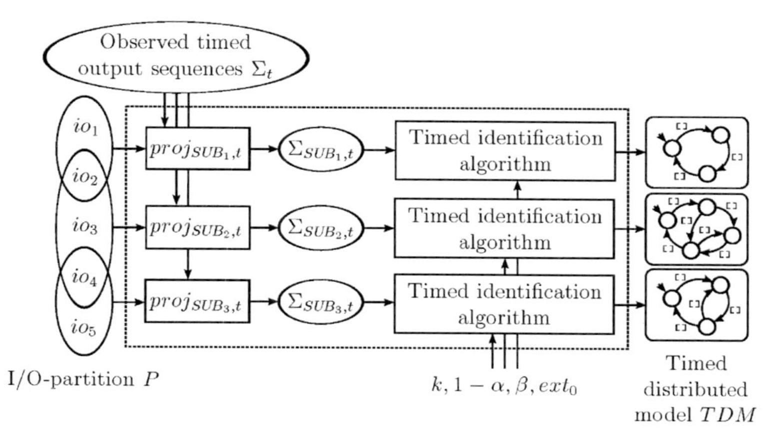

Figure 3.15: Timed distributed identification concept

and the related partial timed output sequences $\Sigma_{SUB_i,t}$, respectively. Finally, a partial automaton $TAAO_i$ is identified for each I/O-subset SUB_i leading to the timed distributed model TDM.

3.5.2 Precision and Completeness Properties

Since the proposed timed identification algorithm is used to identify the partial models of a TDM, the discussed precision and completeness properties hold for each of them, respectively. However, to analyze the precision and completeness of the entire timed distributed model, the combined behavior of all partial automata needs to be considered. Therefore, a *parallel composition*[8] is introduced that builds a timed automaton that represents the combined timed behavior of all partial automata. The composition relies on the parallel composition of timed automata with guards, presented in [Cassandras and Lafortune, 2008]. In order to determine their combined output, the *join output function* is defined. It is used to determine the global output of the model based on the partial automata. The join output function, according to [Roth, 2010], is defined as:

Definition 26 (Join output function). Given two partial automata $TAAO_1$ and $TAAO_2$ and the related partial model outputs $\lambda_{TAAO_1}(x)$ and $\lambda_{TAAO_2}(y)$, with $|\lambda_{TAAO_1}(x)| = |\lambda_{TAAO_2}(y)| = m$, the i-th element of the joined model output is determined as

[8]In literature, parallel composition is also called synchronous composition [Cassandras and Lafortune, 2008].

$$join(\lambda_{TAAO_1}(x),\lambda_{TAAO_2}(y))[i] = \begin{cases} \lambda_{TAAO_1}(x)[i] & \text{if } \lambda_{TAAO_1}(x)[i] = \lambda_{TAAO_2}(y)[i] \\ \lambda_{TAAO_1}(x)[i] & \text{if } (\lambda_{TAAO_1}(x)[i] \neq -) \wedge (\lambda_{TAAO_2}(y)[i] = -) \\ \lambda_{TAAO_2}(y)[i] & \text{if } (\lambda_{TAAO_1}(x)[i] = -) \wedge (\lambda_{TAAO_2}(y)[i] \neq -) \\ c & \text{if } \lambda_{TAAO_1}(x)[i] \neq \lambda_{TAAO_2}(y)[i] \neq - \end{cases} \tag{3.34}$$

and the complete model output as

$$join(\lambda_{TAAO_1}(x), \lambda_{TAAO_2}(y)) = \begin{pmatrix} join(\lambda_{TAAO_1}(x), \lambda_{TAAO_2}(y))[1] \\ join(\lambda_{TAAO_1}(x), \lambda_{TAAO_2}(y))[2] \\ \vdots \\ join(\lambda_{TAAO_1}(x), \lambda_{TAAO_2}(y))[m] \end{pmatrix}. \tag{3.35}$$

The join output function combines the model outputs of two partial automata to the combined output. Therefore, each element of the system output $\lambda_{TAAO_1}(x)$ from $TAAO_1$ is compared with the corresponding element $\lambda_{TAAO_2}(y)$ of the second model $TAAO_2$. If their values are equal, the result is the output of the first model in this element. If one of the outputs is a don't care, then the output of the other partial model is taken. In the case that both elements have defined but differing outputs, a contradiction exists, denoted as c.

Example 3.11 (Join output function). Given two partial automata $TAAO_1$ and $TAAO_2$ with partial model outputs

$$\lambda_{TAAO_1}(x) = \begin{pmatrix} 0 \\ - \\ 0 \end{pmatrix} \text{ and } \lambda_{TAAO_2}(x) = \begin{pmatrix} - \\ 1 \\ 0 \end{pmatrix},$$

the joined output is determined as

$$join\left(\lambda_{TAAO_1}(x), \lambda_{TAAO_2}(x)\right) = \begin{pmatrix} 0 \\ 1 \\ 0 \end{pmatrix}.$$

Definition 27 (Parallel composition of two partial TAAOs). Given two partial timed automata $TAAO_1$ and $TAAO_2$ with state-spaces X_1, X_2, initial states x_{1_0}, x_{2_0}, clocks c_1, c_2, time guard sets TG_1, TG_2, timed transition sets TT_1, TT_2, and state outputs Ω_1, Ω_2. The parallel composed model $TAAO_\|$ of $TAAO_1$ and $TAAO_2$ is determined by the following equations:

$$TAAO_\| = TAAO_1 \parallel TAAO_2 = \left(X_\|, x_{0\|}, C_\|, TG_\|, TT_\|, \Omega_\|, \lambda_\|\right) \tag{3.36}$$

with the state-space

$$X_\| = \{(x_1, x_2) \in X_1 \times X_2 \mid join(\lambda(x_1), \lambda(x_2)) \neq c\}, \tag{3.37}$$

the initial state

$$x_{\|0} = (x_{1_0}, x_{2_0}), \tag{3.38}$$

the set of clocks

$$C_{\|} = \{c_1\} \cup \{c_2\}, \tag{3.39}$$

the set of time guards

$$\begin{aligned} TG_{\|} = &\{guard_{c_1} \in TG_1\} \ \cup \{guard_{c_2} \in TG_2\} \ \cup \\ &\{\{guard_{c_1}, guard_{c_2}\} \mid (guard_{c_1} \in TG_1) \wedge (guard_{c_2} \in TG_2)\}, \end{aligned} \tag{3.40}$$

the set of timed transitions

$$\begin{aligned} TT_{\|} = &\{((x_1, x_2), guard_{c_1}, c_1, (x_1', x_2)) \mid ((x_1, guard_{c_1}, c_1, x_1') \in TT_1) \wedge \\ &join(\lambda(x_1'), \lambda(x_2)) \neq c\} \cup \\ &\{((x_1, x_2), guard_{c_2}, c_2, (x_1, x_2')) \mid ((x_2, guard_{c_2}, c_2, x_2') \in TT_2) \wedge \\ &join(\lambda(x_1), \lambda(x_2')) \neq c\} \cup \\ &\{((x_1, x_2), \{guard_{c_1}, guard_{c_2}\}, \{c_1, c_2\}, (x_1', x_2')) \mid ((x_1, guard_{c_1}, c_1, x_1') \in TT_1) \wedge \\ &((x_2, guard_{c_2}, c_2, x_2') \in TT_2) \wedge (join(\lambda(x_1'), \lambda(x_2')) \neq c) \wedge ((x_1, x_2) \neq (x_1', x_2'))\}, \end{aligned} \tag{3.41}$$

the set of state outputs

$$\Omega_{\|} = \{join(\lambda(u_1), \lambda(u_2)) \mid (u_1 \in \Omega_1) \wedge (u_2 \in \Omega_2)\} \tag{3.42}$$

and the output function

$$\lambda((x_1, x_2)) = join(\lambda(x_1), \lambda(x_2)) \ \forall (x_1, x_2) \in X_{\|}. \tag{3.43}$$

The states of the composed model are determined by the cross product of the partial states X_1 and X_2. Each composed state, in which the resulting joined output does not contain a contradicting element, is a valid state of $TAAO_{\|}$. Equation 3.38 determines the initial states as the combination of the partial initial states. In contrast to a $TAAO$ according to Definition 11, the parallel composed timed automaton $TAAO_{\|}$ models a set of clocks $C_{\|}$ instead of one a single clock c. This set of clocks is the union over the clocks given by the partial automata. The time guards $TG_{\|}$ of $TAAO_{\|}$ are determined by Equation 3.40. A time guard, denoted as $guard_{c_i}$, is defined with respect to the clock $c_i \in C_{\|}$, if $guard_{c_i} \in TG_i$. All time guards, contained in $TG_{\|}$, after the parallel composition, refer to the clocks of the partial automata from which they originate. The set $TG_{\|}$ contains all time guards TG_1, TG_2 and the combined time guards $\{guard_{c_1}, guard_{c_2}\}$. In order to build the set of transitions $TT_{\|}$, three cases need to be considered. The first case is when $TAAO_1$ performs a state transition from x_1 to x_1' while the current state of $TAAO_2$ remains the same. Such a transition is valid if the following combined state (x_1', x_2) has no contradicting output. The resulting transition is given as $((x_1, x_2), guard_{c_1}, c_1, (x_1', x_2))$ with time guard $guard_{c_1}$ and the clock reset $c_1 := 0$. In the second case, similar to the first case, a transition in $TAAO_2$ fires such that $c_2 := 0$ and no transition is executed in

$TAAO_1$. The third case represents the simultaneous execution of a transition in $TAAO_1$ from state x_1 to x'_1 and in $TAAO_2$ from state x_2 to x'_2. Such a transition is valid if the partial transition exists in the partial models, respectively, if the following combined state (x'_1, x'_2) has no contradicting output, and if no self-loop is performed. The combined transition is given by $((x_1, x_2), \{guard_{c_1}, guard_{c_2}\}, \{c_1, c_2\}, (x'_1, x'_2))$ with the set of related partial time guards $guard_{c_1}$ and $guard_{c_2}$ and the clock reset conditions $c_1 := 0$ and $c_2 := 0$. The set of outputs is given by all joined outputs of $TAAO_1$ and $TAAO_2$ for which no contradictions exist and the combined output function is determined by the joined operation of all partial model outputs. An example for this composition will be given during the following discussion on shared I/Os. More than two partial automata can be composed by successively performing the composition operation according to the associative law

$$TAAO_{\|} = TAAO_1 \parallel TAAO_2 \parallel TAAO_3 = (TAAO_1 \parallel TAAO_2) \parallel TAAO_3. \tag{3.44}$$

The timed language modeled by the identified TDM is denoted as $L^n_{Mod,TAAO_{\|},t}$. It represents all timed words that are accepted by the composed automaton $TAAO_{\|}$. Under certain condition, the TDM models the complete original timed language $L^n_{Orig,t}$ of a DES. This completeness refers to the following theorem.

Theorem 3. Given a TDM that is timed identified according to the I/O-partition $P = \{SUB_1, SUB_2, \ldots, SUB_N\}$. If $L^n_{Mod,SUB_i,t} \supseteq L^n_{Orig,SUB_i,t}$ $\forall 1 \leq i \leq N$, then $L^n_{Mod,TAAO_{\|},t} \supseteq L^n_{Orig,t}$ for $n \geq 1$.

The proof is given in Appendix A. In words, if all partial automata based on SUB_i are completely identified such that they represent the original timed language $L^n_{Orig,SUB_i,t}$ of the corresponding subsystem, then the timed language of the composed model $L^n_{Mod,TAAO_{\|},t}$ simulates the timed language of the original system $L^n_{Orig,t}$. As result, it is possible to identify a TDM that is complete with respect to the original timed language of the DES although the observed language L^n_{Obs} of the global system may not converge for any n. Since the TDM is completely identified, the non-reproducible behavior of the model is minimized and the number of false detections during model-based fault diagnosis remains low.

Due to the applied identification algorithm, the presented precision properties hold for each partial automaton. However, for the composed model no precision guarantee can be given, neither for the logical behavior, nor for the time behavior. The reason therefore is that the combination of partial behavior can lead to modeled timed words that have not been observed before. This new generated behavior must not necessarily correspond to the timed language of the original system. However, by proper partitioning of the DES, the additionally generated behavior can be kept to a minimum such that the resulting models are appropriate for fault diagnosis purposes. This will become evident in Chapter 5 by an evalation of the fault detection capabilities using timed distributed models.

3.5.3 Discussion on Shared I/Os

In general, a subsystem of a DES can be characterized by its partial event set $E_i \subseteq E$ with E denoting the event set of the DES. The partial event sets of two subsystems are either pairwise disjoint $E_i \cap E_j = \emptyset$, $i \neq j$ or partially overlapping $E_i \cap E_j \neq \emptyset$. In the first case, the subsystems model only *private events* such that the behavior of all subsystems is independent of each other. Private events are owned exclusively by one subsystem and do only affect its partial behavior. In the second case, the subsystems have common events, so called *shared events* that are contained in both subsets E_i and E_j [Willner and Heymann, 1991]. The behavior of these subsystems depends on each other since shared events are supposed to occur simultaneously. This leads to synchronization effects. A distributed model of a DES is supposed to represent private and shared events. While private events can be executed by the partial models at any time, the shared events restrict the combined behavior of a distributed model. If this dependency is not considered during modeling, it may be possible that a distributed model allows for combined behavior that cannot be produced by the original system due to existing synchronization. In order to model shared events, the I/Os related to these shared events must be contained in the corresponding I/O-subsets. These I/O have been introduced in Definition 21 as *shared I/Os*. If a shared I/O, given as $io_{share} \in IO$, changes its value, all partial models with $io_{share} \in SUB$ perform synchronous state transitions. All other partial models that do not share io_{share} do not perform any state transition because of the 'don't care symbol' in the related partial system output. By modeling shared I/Os during partitioning, the exceeding behavior of timed distributed model TDM can be reduced. This will be shown in the following by means of an example.

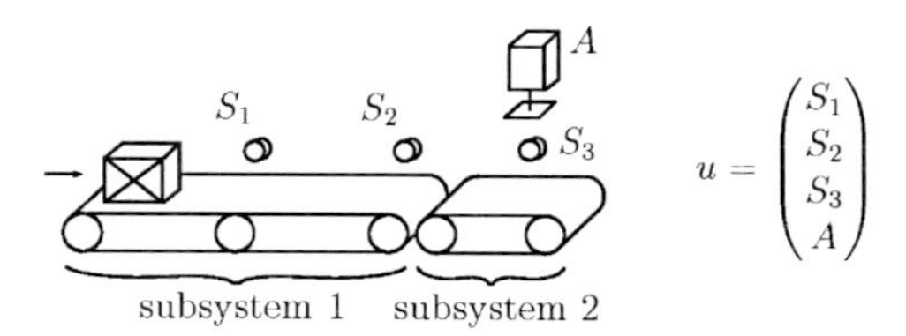

Figure 3.16: Example DES with two concurrent operating subsystems

Example 3.12 (Concurrent operating subsystems). In Figure 3.16, the plant of a concurrent DES is depicted. It consists of two subsystems that are operating independently of each other. The conveyor belts of both subsystems are continuously running with the same speed. A workpiece arrives at random time at subsystem 1 and is then transported to subsystem 2. Only one workpiece can be on a conveyor at the same time. The concurrency is related to the fact that the workpiece positions on the conveyors can vary for different cycles leading to different combinations of sensor and actuator values. The sensors S_1, S_2, and S_3 recognize the presence of a workpiece. If a workpiece arrives at S_3, the cylinder is driven by actuator A while the work piece is moving towards the end of the convertor. The operation of the cylinder is independent of the workpiece's position on the first conveyor. The system output u of the DES is shown in the figure. It is represented as the I/O-vector based on $IO = \{S_1, S_2, S_3, A\}$. In the following, it is assumed that the plant is in a fault state and performs behavior that does not correspond to the fault-free behavior of the system. Therefore, sensor S_3 is assumed to be stuck-off such that its value remains 0, independent of the presence of a workpiece. As a consequence, the arriving of workpieces at the tool cannot be recognized and the cylinder does not operate. Given this fault situation, the following timed output sequence $\sigma_{t,F}$ is observed:

$$\sigma_{t,F} = \left(\begin{pmatrix}0\\0\\0\\0\end{pmatrix}, \bot\right), \left(\begin{pmatrix}1\\0\\0\\0\end{pmatrix}, 1422\right), \left(\begin{pmatrix}0\\0\\0\\0\end{pmatrix}, 5\right), \left(\begin{pmatrix}0\\1\\0\\0\end{pmatrix}, 12\right), \left(\begin{pmatrix}0\\0\\0\\0\end{pmatrix}, 6\right), \left(\begin{pmatrix}1\\0\\0\\0\end{pmatrix}, 20\right), \left(\begin{pmatrix}0\\0\\0\\0\end{pmatrix}, 4\right), \left(\begin{pmatrix}0\\1\\0\\0\end{pmatrix}, 12\right), \ldots$$

Workpieces are periodically arriving at the first conveyor and then, they are transported to the next subsystem. This is indicated by the changing values of S_1 and S_2 with ongoing number of observations. Sensor S_3 does not change its value due to the stuck-off fault, hence actuator A does not start operation and its value remains 0 in all system outputs of the observed sequence. A model of the considered system that represents the fault-free behavior is supposed to be unable to reproduce this faulty output sequence $\sigma_{t,F}$. In Figure 3.17, an identified TDM based on the disjoint I/O-partition

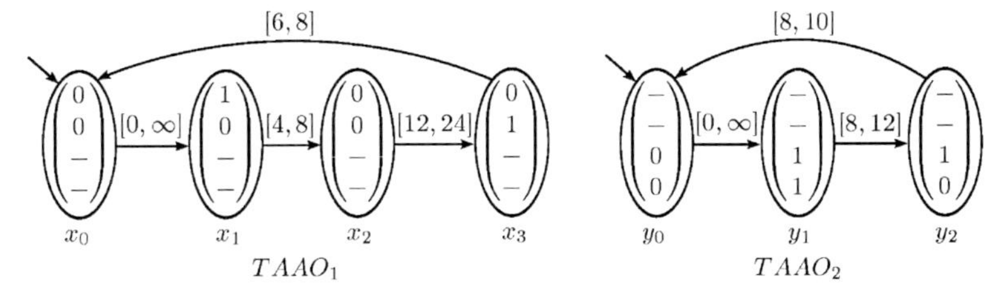

Figure 3.17: Partial automata based on P_{disj} without shared I/Os

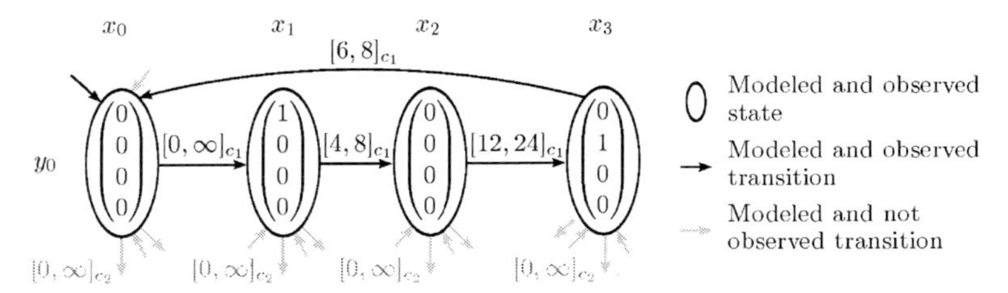

Figure 3.18: State trajectory upon $\sigma_{t,F}$ in the composed model based on P_{disj}

$$P_{disj} = \{\{S_1, S_2\}, \{S_3, A\}\}$$

with $SUB_1 \cap SUB_2 = \emptyset$, is given. The partial automaton for $SUB_1 = \{S_1, S_2\}$ models the behavior of the first subsystem while $TAAO_2$ models the behavior of the second subsystem, referring to $SUB_2 = \{S_3, A\}$. The time guards from the initial states $guard(x_0, x_1)$ and $guard(y_0, y_1)$ are maximal permissive since the random arriving of workpieces prevents from the identification of meaningful time guards. The combined behavior of the partial automata are determined by the parallel composition, according to Definition 27. Figure 3.18 shows the relevant section of the composed automaton. The entire composed automaton is given in the Appendix B. The state trajectory based on $\sigma_{t,F}$ is drawn with black lines while modeled but non-observed transitions are gray shaded. The composed automaton is initialized in state (x_0, y_0) in which the first observed system output is reproduced. With ongoing observations, the state trajectory

$$(x_0, y_0) \to (x_1, y_0) \to (x_2, y_0) \to (x_3, y_0) \to (x_0, y_0) \to (x_1, y_0) \to (x_2, y_0) \to (x_3, y_0) \to \dots$$

is produced. As a result, the given observed faulty output sequence $\sigma_{t,F}$ can be reproduced by the composed automaton model that represents the combined behavior of the TDM. Although the TDM is identified based on observations of the fault-free system behavior, it models exceeding timed behavior that corresponds to faulty DES behavior. A fault diagnosis implementation using this model is not able to detect the stuck-off fault of S_3. Another TDM of the considered DES is given with the I/O-partition

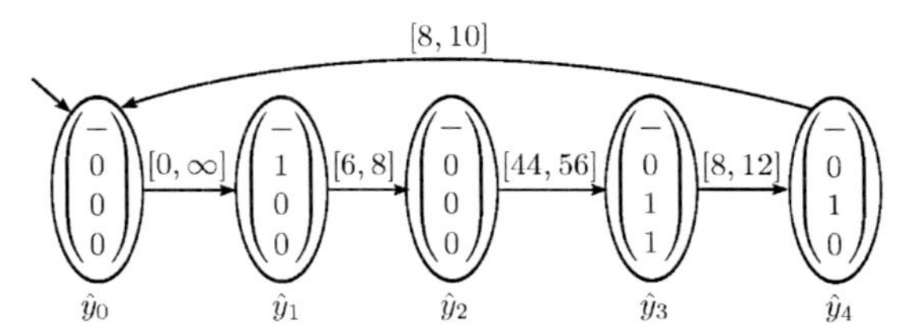

Figure 3.19: Partial automaton $\widehat{TAAO}_2$ based on P_{share} with shared I/Os

$$P_{share} = \{\{S_1, S_2\}, \{S_2, S_3, A\}\}$$

with the shared I/O S_2. The partial automaton $TAAO_1$ remains the same as for the previous I/O-partition given in Figure 3.17, since $SUB_1 = \{S_1, S_2\}$ is equivalently defined. The new partial automaton $\widehat{TAAO}_2$, based on the modified I/O-subset $\widehat{SUB_2} = \{S_2, S_3, A\}$, is depicted in Figure 3.19. If the shared I/O S_2 changes its value, both partial automata are forced to synchronously perform a state transition. This can be seen in the section of the composed automaton given in Figure 3.20. The complete composed automaton is again given in Appendix B. Consider, for instance, the transition from the combined state $(x_2, \hat{y}_0)$ to state $(x_3, \hat{y}_1)$. Due to the changing value of S_2, both partial automata simultaneously perform a transition. The local execution of a transition by a single partial automaton is not permitted. This would lead to contradicting outputs, which is not allowed for the combined behavior. Note that if more than one partial automaton performs a state transition at the same time, the time behavior, given by the local clocks, must be consistent with the combined time guards related to the executed transitions. In the figure of the composed automaton, states and transition that are part of the state trajectory for $\sigma_{t,F}$ and marked again with black color while modeled but not

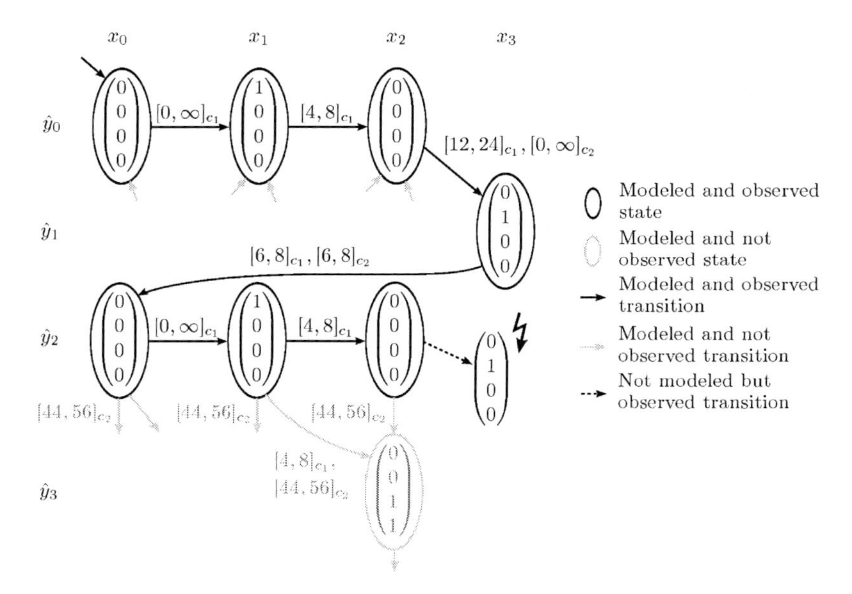

Figure 3.20: State trajectory upon $\sigma_{t,F}$ in the composed model based on P_{share}

observed states and transitions are gray shaded. Furthermore, the dotted transition corresponds to observed behavior that cannot be reproduced by the model. Based on SUB_1 and $\widehat{SUB_2}$, the state trajectory in the composed automaton for $\sigma_{t,F}$ results in

$$(x_0, \hat{y}_0) \rightarrow (x_1, \hat{y}_0) \rightarrow (x_2, \hat{y}_0) \rightarrow (x_3, \hat{y}_1) \rightarrow (x_0, \hat{y}_2) \rightarrow (x_1, \hat{y}_2) \rightarrow (x_2, \hat{y}_2).$$

The new observation made after activation of $(x_2, \hat{y}_2)$ cannot be reproduced by the composed model. Hence, the faulty behavior combination of the subsystems can be recognized. A fault diagnosis implementation using this improved model is able to detect the stuck-off fault of S_3 after a number of made observations. By modeling S_2 as a shared I/O of the two subsystems, the exceeding behavior of a TDM can be reduced. Shared I/Os are necessary to represent the restrictive dependencies between subsystems in the model. As a consequence, the I/O-partition for a timed distributed fault diagnosis model has to represent the shared I/Os in order to ensure model precision.

3.6 Identification of Timed Distributed BMS Models

3.6.1 Data Collection

The application of the presented timed identification approach for timed distributed models to the BMS is presented in the following. The evaluation system has been described

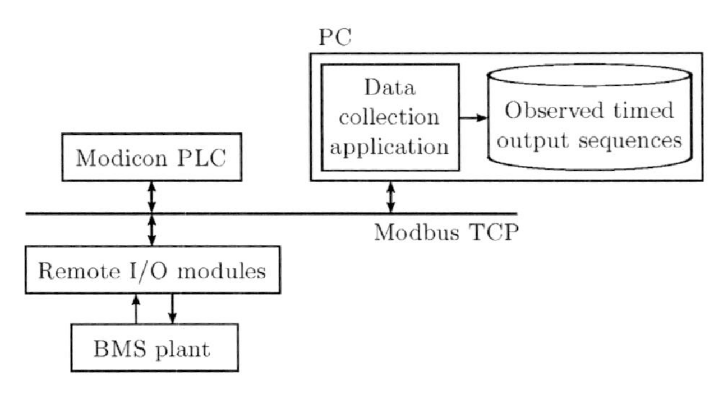

Figure 3.21: Data collection implementation of the BMS

in the introduction chapter of this thesis. For identifying a fault-free model, the BMS was observed online during its fault-free operation. Therefore, a data transmission procedure was implemented on the PLC and a data collection application on a PC. The implementation concept is depicted in Figure 3.21. PLC and plant are communicating using a Modbus TCP protocol and the sensors and actuators of the plant are accessed via remote I/O modules. These modules act as gateways between the hard-wired sensor and actuator signals and the Modbus. The timed system outputs, generated by the BMS, are periodically transmitted by the PLC to the connected PC. The PC records the timed system outputs and stores them as sequences in an attached data-base. This data-base finally contains the observed timed output sequences, which can be used for automatic modeling of the BMS. The PLC procedures for data transmission were implemented such that the operation of the BMS was not affected. This meets the requirement for passive observation.

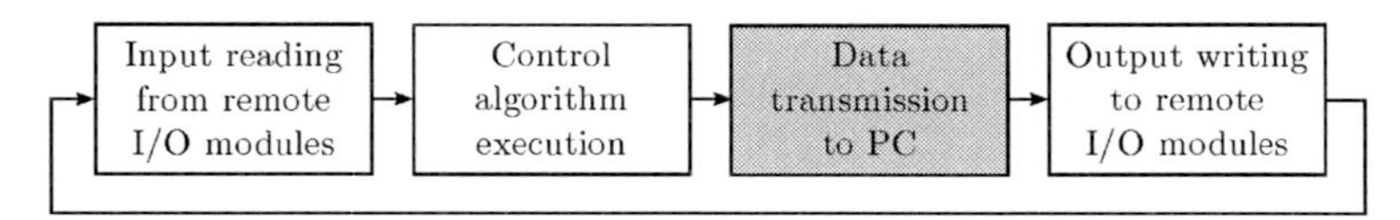

Figure 3.22: PLC cycle with data transmission

In Figure 3.22, the extended PLC operation cycle is illustrated. The observation procedures were implemented in the stage "data transmission to PC". After the execution of the control algorithm, the determined values of all I/Os are compared with the values of the former PLC cycle. If the value of one or more I/Os has changed, the current system output $u(j)$ and the output time spans $\delta(j)$ are recorded by the PLC. These observations are then periodically sent to the PC where they are stored as observed timed output sequences.

3.6.2 Timed Distributed Identification

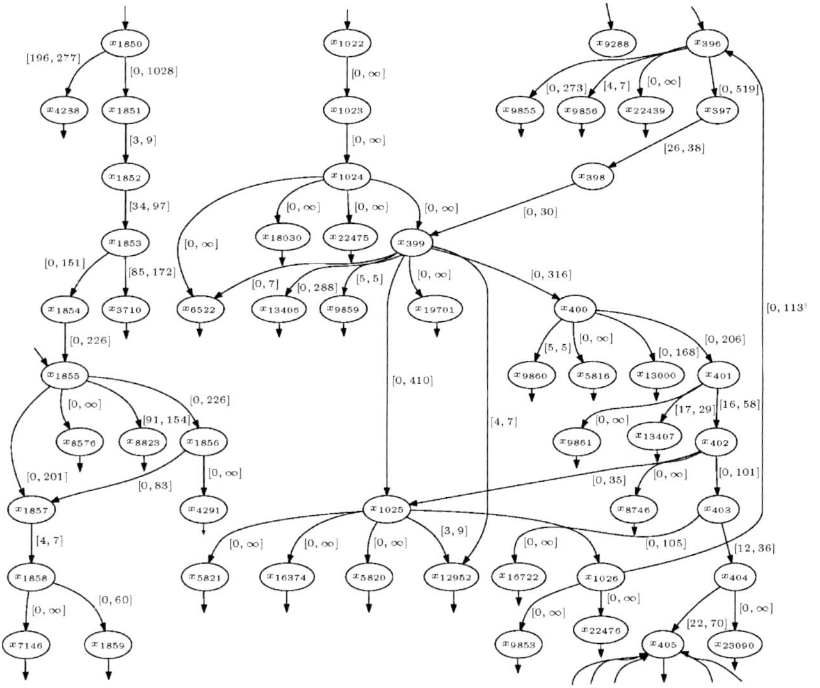

Figure 3.23: Section of the identified monolithic *TAAO*

The timed identification of the monolithic model was performed based on 50 observed timed output sequences Σ_t. The logical identification parameter was selected as $k = 1$ and the time identification parameter as $1 - \alpha = 0.99$, $\beta = 0.95$, and $ext_0 = 1.0$. Given $1 - \alpha$ and β, the minimum number of transition time spans observations results in $v_0 = 1720$ according to Equation 3.23. A section of the identified monolithic *TAAO* is illustrated in Figure 3.23. The automaton contains $|X| = 23662$ states and $|TT| = 30845$ timed transitions. The identification took in total 535 s with a desktop PC.[9]

In order to evaluate the timed identification result, characteristic values are determined for the identified model. First, the *mean and maximum number of observed time spans* available for time identification are considered. The mean number of observed time spans

[9] Intel(R) Core(TM) i3-530 2.94 GHz.

of all transitions in all identified automata is determined as

$$\bar{v} = \frac{1}{N} \sum_{TAAO_i \in TDM} \left(\frac{1}{|TT_{TAAO_i}|} \sum_{(x,guard,c,x') \in TT_{TAAO_i}} |\Delta_{Obs}^{(x,x')}| \right) \tag{3.45}$$

and the maximum number of observed time spans of all transitions in all partial automata by equation

$$v_{max} = \max_{TAAO_i \in TDM} \left(\max_{(x,guard,c,x') \in TT_{TAAO_i}} |\Delta_{Obs}^{(x,x')}| \right) \tag{3.46}$$

with $|\Delta_{Obs}^{(x,x')}|$ as the number of time spans contained in the observed time spans sequence $\Delta_{Obs}^{(x,x')}$.[10] With respect to the required number of observed time spans v_0, $\bar{v}$ and v_{max} indicate whether the bounds of the time span sets can be considered as completely converged $\min\left(\Delta_{Obs}^{(x,x')}\right) \approx \min\left(\Delta_{Orig}^{(x,x')}\right)$ and $\max\left(\Delta_{Obs}^{(x,x')}\right) \approx \max\left(\Delta_{Orig}^{(x,x')}\right)$. If not, the tolerance intervals are extended to reduce the non-reproducible time behavior. This leads to another characteristic given by the time period that all tolerance intervals are extended by the identification algorithm with respect to the observed time span sequences. This period is denoted as *generalized time period* θ in the following. Only transitions are considered for which $v > 1$ holds. For a transition $(x, guard, c, x') \in TT$, the generalized time period is determined by the following equation

$$\theta(x,x') = \max\left(guard(x,x')\right) - \min\left(guard(x,x')\right) - \left(\max\left(\Delta_{Obs}^{(x,x')}\right) - \min\left(\Delta_{Obs}^{(x,x')}\right)\right) \tag{3.47}$$

and for all transitions in all automata of a TDM as

$$\theta = \sum_{TAAO_i \in TDM} \sum_{(x,guard,c,x') \in TT_{TAAO_i}} \theta(x,x'). \tag{3.48}$$

The generalized time period represents the tolerance added to the minimum and maximum observed time spans, respectively. If $|\Delta_{Obs}^{(x,x')}| \geq v_0$ or $ext_0 = 0$, then $\theta(x,x') = 0$. Minimizing the generalized time period $\theta(x,x')$ also minimizes the exceeding time behavior $\Delta_{EXC}^{(x,x')}$. The last considered characteristic of a temporally identified model is given by the *ratio of time-constrained transitions*. Time-constrained transitions have time guards $guard$ attached that do not correspond to the maximum permissive ones, i.e. $guard \neq [0, \infty]$. The ratio for the i-th partial automaton is determined as

$$\gamma_{TAAO_i} = \frac{1}{|TT_{TAAO_i}|} |\{(x, guard, c, x') \in TT_{TAAO_i} \mid (guard \neq [0,\infty])\}| \tag{3.49}$$

with the timed transitions TT_{TAAO_i} related to $TAAO_i$ and for all automata of a TDM as

$$\gamma = \sum_{TAAO_i \in TDM} \gamma_{TAAO_i}. \tag{3.50}$$

[10]For the monolithic model, $TDM = \{TAAO\}$ holds.

Mean number of observed time spans $\bar{v}$	Maximum number of observed time spans v_{max}	Ratio of time-constrained transitions γ	Generalized time period θ
3.27	203	42.39%	760.51 s

Table 3.1: Characteristic values of the identified monolithic *TAAO*

Time guards that are maximum permissive do not restrict the time behavior of the model. They allow for a maximum amount of exceeding behavior $\Delta_{EXC}^{(x,x')}$. Therefore, the number of maximum permissive time guards should be as low as possible compared to the absolute number of transitions.

The determined characteristic values for the identified monolithic *TAAO* are summarized in Table 3.1. The results of the number of observed time spans show that the data-base for time identification was poor with respect to the required number of observed time spans $v_0 = 1720$. In average, only 3.27 time span observations were available for each transition $(x, guard, c, x')$ to determine a tolerance interval. This is also indicated by the maximum number of observed time spans v_{max}. Furthermore, many transitions exist that were only observed once, leading to a large amount of transitions that are unconstrained. Only 42.39% of all transitions have 'truly' identified time guards. The generalized time period θ indicates that tolerance intervals were extended by 760.51 s in sum.

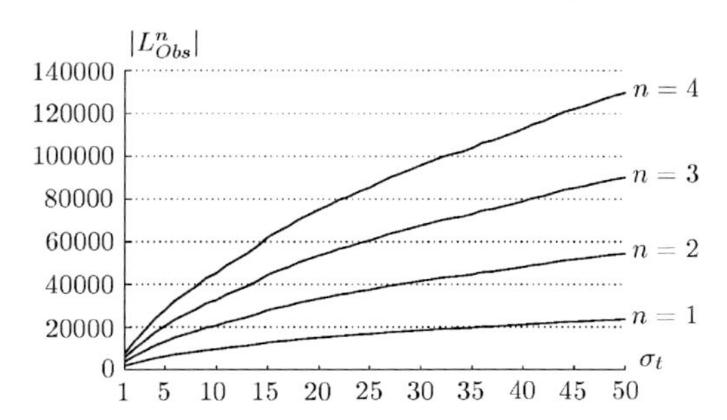

Figure 3.24: Observed language growth of the monolith

The BMS is a closed-loop DES with a high degree of concurrency. It consists of several subsystems while some are synchronized with each other and others are independently operating. The two pneumatic presses, for instance, are two independent operating subsystems. They start operating when they receive a workpiece from the attached conveyor, independent of the operation of the other press, respectively. In contrast to that, the press and the attached conveyor operate concurrently too, but both machines are synchronized at some time instance such that a workpiece can be handed over from one machine to the other. In order to obtain a complete monolithic timed model of the BMS, a very high number of observations is necessary to capture all different behavior combination of the concurrent operating subsystems. In Figure 3.24, the language growth of the observed

i	1	2	3	4	5	6	7	8	9	10
$\lvert X_{TAAO_i}\rvert$	82	33	28	17	18	56	53	17	8	12
$\lvert TT_{TAAO_i}\rvert$	102	56	43	28	18	75	76	32	16	27

Table 3.2: Number of states and transitions for each $TAAO_i$ of the TDM

Mean number of observed time spans $\bar{v}$	Maximum number of observed time spans v_{max}	Ratio of time-constrained transitions γ	Generalized time period θ
346.7	2244	99.58%	481.172 s

Table 3.3: Characteristic values of the identified TDM

logical languages is shown, based on the 50 observed timed output sequences Σ_t. It can be seen that none of observed languages L^n_{Obs} converges for any $n > 1$. Consequently, it is not possible to identify a complete monolithic timed model based on the available data. On the contrary, the identified monolithic model, shown in Figure 3.23, must be considered as incomplete because of the non-convergence. Using this model for fault detection purposes will lead to a high number of false detections. This will be demonstrated in Chapter 5 by model validation.

In order to completely model the original behavior of the BMS, based on the given set of observed timed outputs sequences Σ_t, a timed distributed model was identified. Therefore, a given I/O-partition P was applied. The partition consists of $N = 10$ I/O-subsets representing the 73 I/Os of the BMS. 21 of the I/Os are modeled as shared I/Os, they are contained in more than one I/O-subset. Given the observed partial timed output sequences, the growth of the observed languages for all I/O-subsets $SUB_i \in P$ are depicted in Figure 3.25. One can see that for each SUB_i it exists a $n > 1$ such that L^n_{Obs,SUB_i} can be considered as converged after a number of observations. This holds, for instance, for $n = 2$ in all I/O-subsets. Hence, it can be concluded that $L^n_{Obs,SUB_i} \approx L^n_{Orig,SUB_i}, \forall 1 \leq i \leq N$ and all partial automata $TAAO_i$ can be logically completely identified. The timed identification of the 10 partial automata, based on the identification parameters given for the monolithic identification, took 24 s in total. The number of states and transitions obtained for each partial model $TAAO_i \in TDM$ are listed in Table 3.2. In sum, the TDM consists of 324 states and 473 transitions. These numbers are significantly smaller than for the monolithic model, which requires 23662 states and 30845 transitions in order to model the same observed behavior. As a result, the state-space explosion can be avoided by using a distributed model for the BMS. In Table 3.3, the determined characteristic values of the timed identification are given. The mean number of observed time spans for each transition is determined as 346.7, which is smaller than $v_0 = 1720$. Most time guards of the timed distributed model were extended to obtain a complete model of the time system behavior. However, there exist transitions, indicated by $v_{max} = 2244$, which were completely observed with respect to v_0. The distributed identification led to a model in which 99.58% of all transitions are time-constrained. This is very appropriate for the detection of time related faults since the evaluation of the time behavior requires time

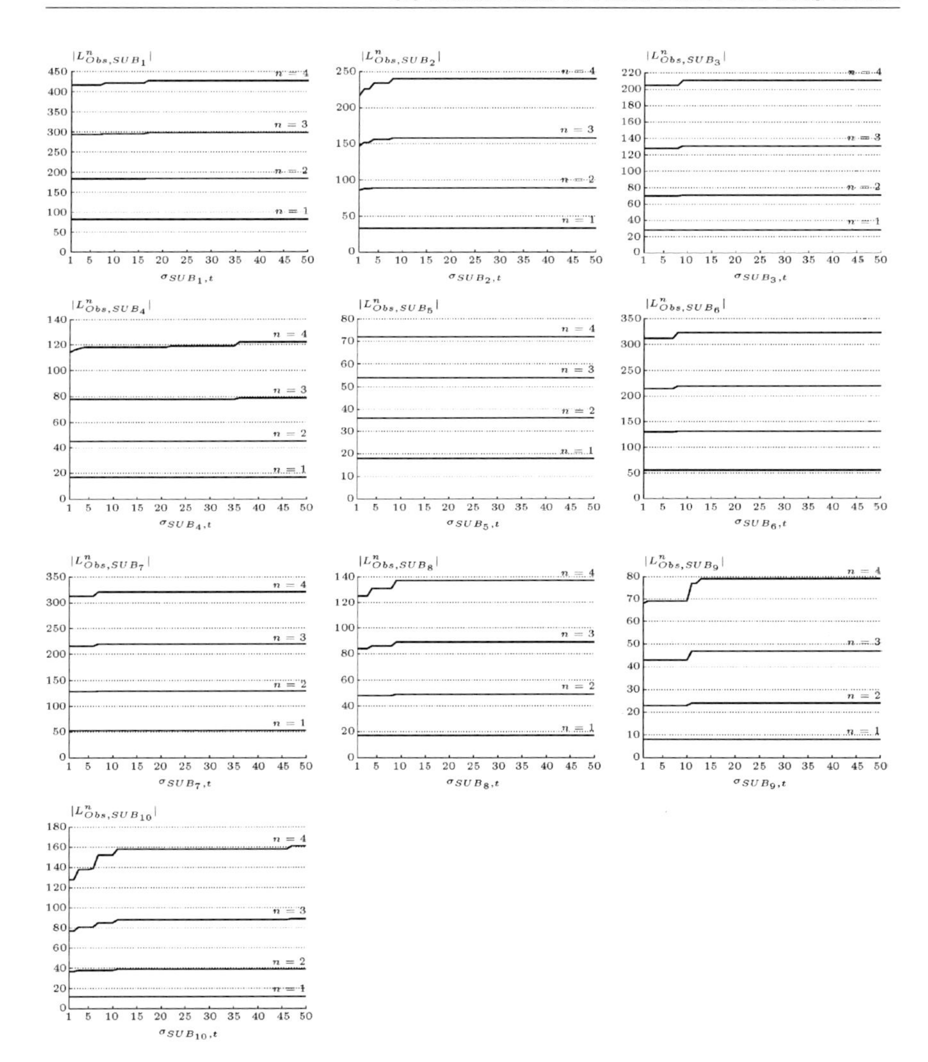

Figure 3.25: Observed language growth for all I/O-subsets of the I/O-partition P

guards that do not correspond to the maximum permissive ones. The generalized time period set is determined as $\theta = 481.172\,\mathrm{s}$ for the given identification parameters. This is comparable less time than determined for the monolithic model.

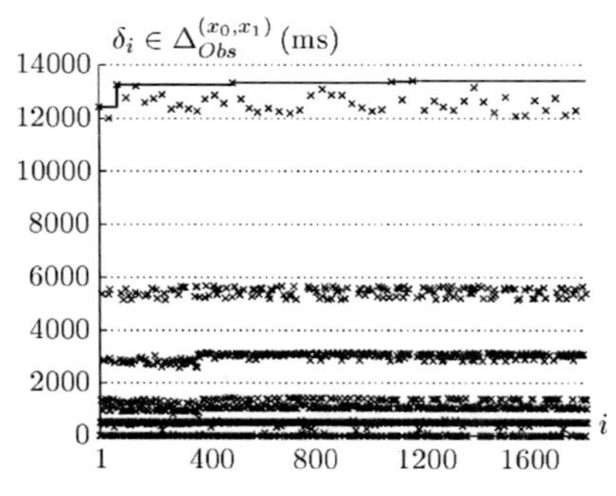

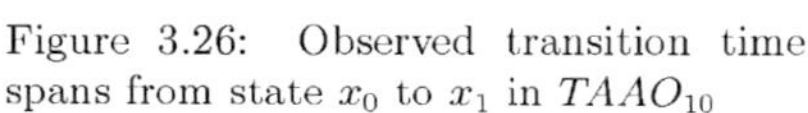
Figure 3.26: Observed transition time spans from state x_0 to x_1 in $TAAO_{10}$

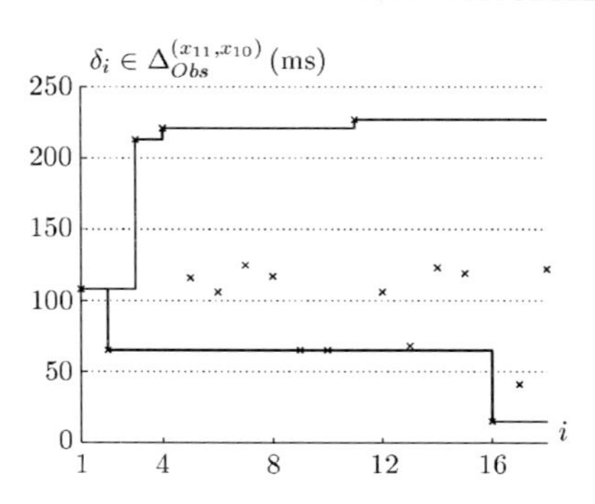

Figure 3.27: Observed transition time spans from state x_{11} to x_{10} in $TAAO_{10}$

In the Figures 3.26 and 3.27, the observed time span sequences $\Delta_{Obs}^{(x,x')}$ for the two example transitions of the partial automaton $TAAO$ are depicted. The diagram in Figure 3.26 shows the observed timed spans $\Delta_{Obs}^{(x_0,x_1)}$ for the transition from state x_0 to x_1. The transition has been observed $|\Delta_{Obs}^{(x_0,x_1)}| = 1889$ times, hence $|\Delta_{Obs}^{(x_0,x_1)}| \geq v_0$ holds. The observed minimum and maximum bounds $\min \Delta_{Obs}^{(x_0,x_1)}$ and $\max \Delta_{Obs}^{(x_0,x_1)}$ are highlighted by a solid line. Since $|\Delta_{Obs}^{(x_0,x_1)}| \geq v_0$, the bounds of the observed time span sequence can be considered as converged to the bounds of the corresponding original, unknown time span set. This is visible in the diagram by the fact that after approximately $v = 1100$ no time spans are observed beyond $\min(\Delta_{Obs}^{(x_0,x_1)})$ and $\max(\Delta_{Obs}^{(x_0,x_1)})$, respectively. These observed time bounds have not been extended during time identification. In contrast to that, Figure 3.27 shows the observed timed spans for the transition from state x_{11} to x_{10} with $|\Delta_{Obs}^{(x_0,x_1)}| < v_0$. Due to the small number of available observations, the minimum and maximum bounds cannot be considered as converged. In future, there will likely be time spans observed that are larger than the maximum and smaller than the minimum observes ones. In order to identify a complete time guard for this transition, the observed bounds have been extended during identification, accordingly.

In general, the number of observed time spans of several transitions in an automaton differs from each other. This is illustrated in Figure 3.28 by means of the identified partial automaton $TAAO_4$. The automaton is drawn as a weighted graph such that the weight of the edge corresponds to the number of observed time spans v. The transition from state x_5 to state x_6 was observed most often $v = 900$ while transition from x_{16} to x_{15} was observed fewest, in total $v = 12$ times. The figure shows that the data basis for the time guard identification may not be homogeneous within an automaton, which poses differing conditions for the time guards identification. The proposed tolerance extension approach considers this fact by adapting to tolerance according to the number of made observations.

The next diagrams illustrate the effect of the spread of observed time spans sequences $(\max(\Delta_{Obs}^{(x,x')}) - \min(\Delta_{Obs}^{(x,x')}))$ and of the number of observed time spans v on the time

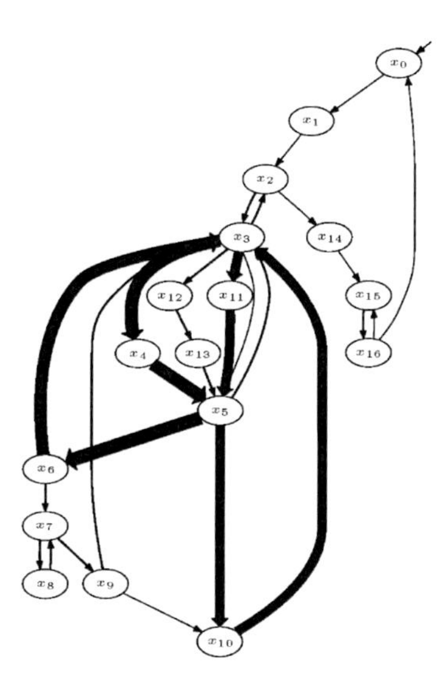

Figure 3.28: Partial automaton $TAAO_4$ with edge weight v

guard determination. The extension parameter was selected as $ext_0 = 1.0$ for all identified time guards. In the first case, two transitions with different spread are considered. In Figures 3.29 and 3.30, the two observed time span sequences $\Delta_{Obs,TAAO_1}^{(x_5,x_6)}$ and $\Delta_{Obs,TAAO_1}^{(x_9,x_4)}$ and the corresponding identified time guards $guard(x_5, x_6)$ and $guard(x_9, x_4)$ are shown. The number of observed time spans $v = 294$ is equal for both observed time span sequences. Hence, the effect of v on the determined tolerance extension $ext(\Delta_{Obs}^{(x,x')})$ was the same in both cases. The different time guard extension refers to the different spread $(\max(\Delta_{Obs}^{(x,x')}) - \min(\Delta_{Obs}^{(x,x')}))$ of the observed transition time spans $\Delta_{Obs}^{(x,x')}$. Since the spread of $\Delta_{Obs,TAAO_1}^{(x_9,x_4)}$ was larger than for $\Delta_{Obs,TAAO_1}^{(x_5,x_6)}$, $guard(x_9, x_4)$ was identified with greater tolerance than $guard(x_5, x_6)$. The generalization time periods can be determined as $\theta(x_9, x_4) = 444\,\text{ms}$ and $\theta(x_5, x_6) = 156\,\text{ms}$. In the next two diagrams, given in Figures 3.31 and 3.32, the effect of the number of observed time spans v on the tolerance extension is investigated. Therefore, the two observed time span sequences $\Delta_{Obs,TAAO_8}^{(x_2,x_1)}$ and $\Delta_{Obs,TAAO_6}^{(y_7,y_{12})}$ and the identified time guards $guard(x_2, x_1)$ and $guard(y_7, y_{12})$ are considered. Both observed time span sequences had approximately the same spread $(\max(\Delta_{Obs,TAAO_8}^{(x_2,x_1)}) - \min(\Delta_{Obs,TAAO_8}^{(x_2,x_1)})) = 93\,ms$ and $(\max(\Delta_{Obs,TAAO_6}^{(y_7,y_{12})}) - \min(\Delta_{Obs,TAAO_6}^{(y_7,y_{12})})) = 100\,ms$ but different number of observations $v = 1449$ for $\Delta_{Obs,TAAO_8}^{(x_2,x_1)}$ and $v = 123$ for $\Delta_{Obs,TAAO_6}^{(y_7,y_{12})}$. The generalized time periods for the transitions are determined as $\theta(x_2, x_1) = 30\,\text{ms}$ and as $\theta(y_7, y_{12}) = 168\,\text{ms}$. As a result, the tolerance interval for $\Delta_{Obs,TAAO_6}^{(y_7,y_{12})}$ was extended comparable more than for $\Delta_{Obs,TAAO_8}^{(x_2,x_1)}$ due to the smaller number of available observations.

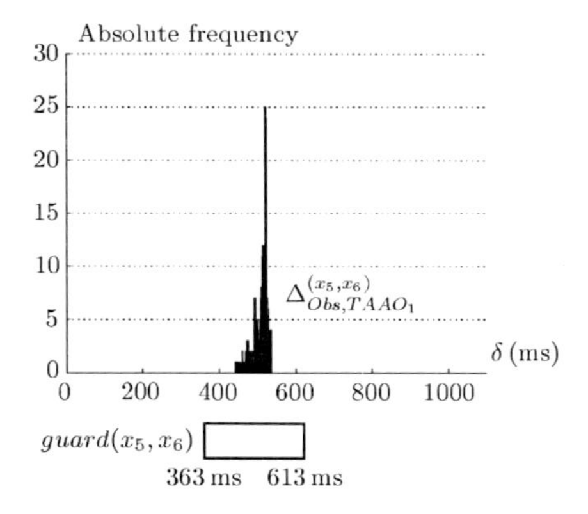

Figure 3.29: Observed time spans and identified time guard for transition from state x_5 to x_6 in $TAAO_1$

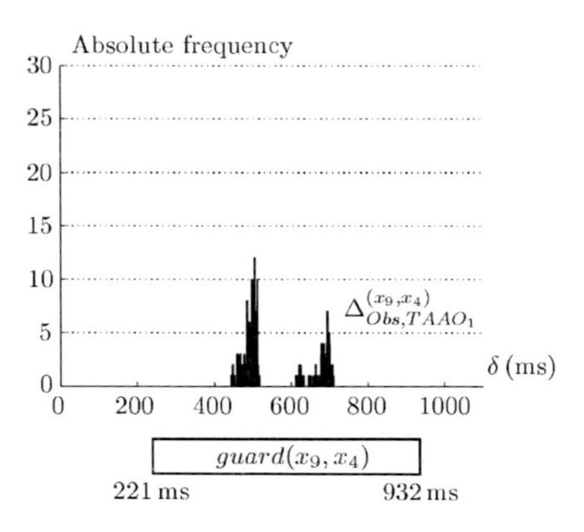

Figure 3.30: Observed time spans and identified time guard for transition from state x_9 to x_4 in $TAAO_1$

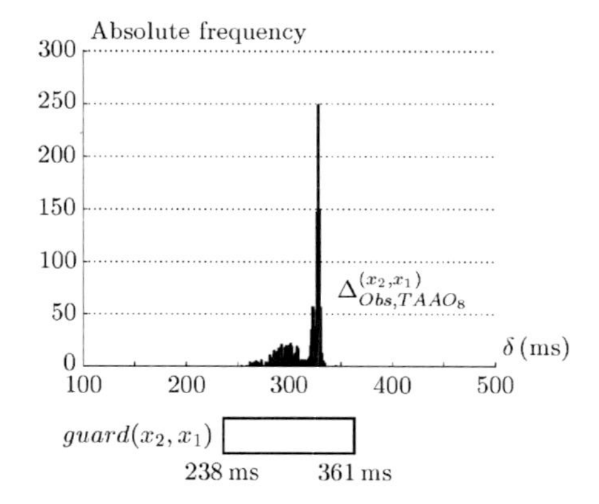

Figure 3.31: Observed time spans and identified time guard for transition from state x_2 to x_1 in $TAAO_8$

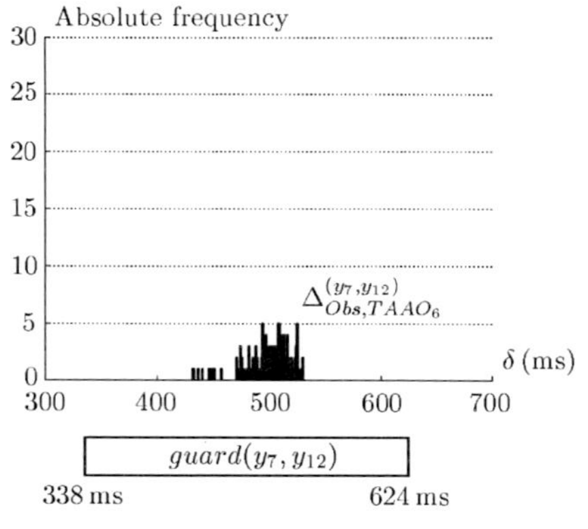

Figure 3.32: Observed time spans and identified time guard for transition from state y_7 to y_{12} in $TAAO_6$

Chapter 4

Partitioning of DES Models

4.1 Preliminaries

In Chapter 3, a distributed modeling approach for concurrent closed-loop DES has been presented. It has been shown that one condition to identify a complete monolithic timed model is to completely observe the logical behavior of the global system $L^n_{Obs} = L^n_{Orig}$. Since concurrent systems can often perform a large amount of behavior due to the combined operation of the subsystems, an impractical large number of observations is required in order to satisfy this condition. By using a distributed modeling approach, concurrent operating subsystems are modeled by partial automata that can be completely identified based on a comparable smaller number of observed output sequences. The question that arises in this context is: How to determine an appropriate I/O-partition $P = \{SUB_1, SUB_2, \ldots, SUB_N\}$ of a DES for a given set of observations Σ? In particular, two issues have to be considered: First, how many I/O-subsets N are required and second, how to distribute the I/Os of the DES among the N I/O-subsets such that all I/Os are considered. If no expert knowledge is available, these issues have to be solved data-based by *automatic partitioning*.

Partitioning of a closed-loop DES is a combinatorial problem whose complexity depends on the number of controller I/Os. Given m I/Os, the number of I/O-partitions with disjoint subsets can be determined by the *Bell number*. The number is derived by the Equation given in [Aigner, 2007]:

$$B(m) = \sum_{N=0}^{m} S(m, N) \tag{4.1}$$

with N the number of I/O-subsets and $S(m, N)$ the Stirling number of the second kind given in [Riordan, 2002] as

$$S(m, N) = \frac{1}{N!} \sum_{j=0}^{N} \binom{N}{j} (-1)^j (N-j)^m. \tag{4.2}$$

For the BMS with $m = 73$, $B(73) = 2.15 \cdot 10^{77}$ is obtained. The combinations referring to $B(m)$ do not consider I/O-partitions with shared I/Os. If shared I/Os are allowed, the number of possible combinations is even larger. Because of the large solutions space, it is impractical to simply test all possible I/O-partitions to find appropriate ones for distributed modeling. Instead, advanced methods are required that test only an exclusive selection of I/O-partitions that likely lead to a distributed model with desired properties.

The automatic determination of models for concurrent systems is pursued in different works in literature. In [Cook et al., 2004], [Maruster et al., 2003], and [Estrada-Vargas, 2013], partitioning is performed using probabilistic data mining approaches. The idea is to extract information from the observed output sequences to discover sequential and concurrent relationships between subsystems. This information is used during identification to build models with distributed structure. Other works as [Medeiros et al., 2007] and [Roth, 2010] interpret partitioning as an optimization problem. They search for I/O-partitions that are globally optimal with respect to a proper defined optimization criterion. In general, this partitioning procedure leads to complex optimization problems that can be hardly solved for highly concurrent DES, especially in the case when distributed models with shared I/Os are desired to reduce the modeled exceeding behavior. These approaches are working well for systems with a rather low degree of concurrency. A comprehensive overview about existing partitioning approaches is given in Chapter 6.

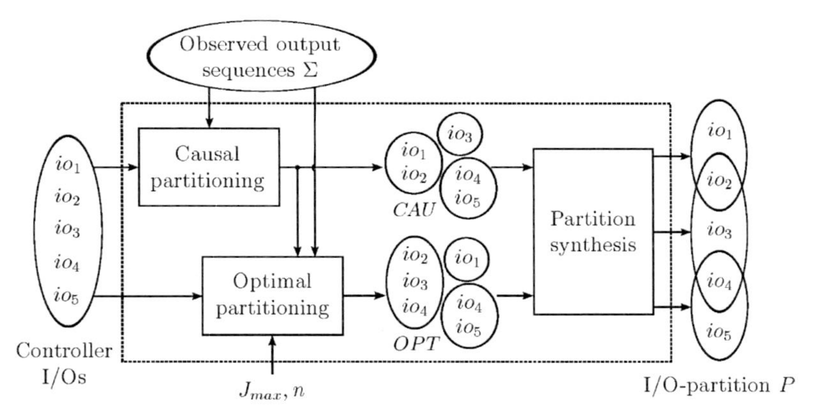

Figure 4.1: Overview of the proposed partitioning approach

In this chapter, an automatic partitioning approach is presented that divides the global partitioning problem into several local problems and determines an I/O-partition P by synthesis of the local solutions. The local problems are considered as finding possible I/O-subsets $\forall io_i \in IO$. Therefore, the two approaches *causal* and *optimal partitioning* are presented, which both rely on the logical fault-free system behavior given by the observed output sequences Σ. Expert knowledge or additional documentation is not required for partitioning. The solution partition P is composed by a selection of the generated I/O-subsets given by $CAU \cup OPT$, with CAU denoting the I/O-subsets generated by causal partitioning and OPT denoting the I/O-subsets generated by the optimal partitioning approach. The presented partitioning approach is appropriate for large systems with a high degree of concurrency and generates I/O-partitions with shared I/Os. An overview about the approach is illustrated in Figure 4.1.

4.2 Causal Partitioning

4.2.1 Distance and Causality

The idea of causal partitioning is to determine I/O-subsets in which all I/Os have a *causal relationship* to each other. I/Os are considered as causally related if the related partial output sequences are equal, i.e. the generated logical behavior is the same in all evolutions of the DES. Causally related I/Os are advantageous for identifying a partial model since their behavior can be typically completely observed by means of few output sequences. An example for a system with causally related I/Os is given with the example manufacturing system in Figure 3.16, which has been introduced for the discussion of shared I/Os. The partial behavior produced by the I/Os of the subsystems, respectively, is the same in all system evolutions. In subsystem 1, an arriving workpiece is always recognized by sensor S_1 first and then by sensor S_2. This behavior is the same for all production cycles. The same holds for the I/Os of subsystem 2. In each system evolution, the workpiece is recognized by S_3 first and then actuator A starts operating. Although the individual logical behavior of the subsystems is strictly sequential, due to disturbances and different timings their combined behavior may become very large. It is presented in the following how the causally related I/Os can be determined automatically based on observations of the fault-free logical DES behavior. Therefore, the observed output sequences need to be redefined.

Definition 28 (Observed output sequences). The set of observed output sequences of the DES is given as $\Sigma = \{\sigma_1, \sigma_2, \ldots, \sigma_h, \ldots, \sigma_p\}$ and the h-th sequence is given as $\sigma_h = (u_h(1), u_h(2), \ldots, u_h(l_h))$.

The observed time behavior of a DES, given by $t(j)$, is not considered for partitioning. All notations used in the following refer to the logical behavior of the system, exclusively. The determination of causal related I/Os relies on a comparison of *partial output sequences.* These sequences are defined as:

Definition 29 (Observed partial output sequences). Given an I/O-subset $SUB \subseteq IO$ of a DES, the observed partial output sequences are given as

$$\Sigma_{SUB} = \{\sigma_{SUB,1}, \sigma_{SUB,2}, \ldots, \sigma_{SUB,h}, \ldots, \sigma_{SUB,p}\} \quad (4.3)$$

with the h-th sequence given as $\sigma_{SUB,h} = (u_{SUB,h}(1), u_{SUB,h}(2), \ldots, u_{SUB,h}(l_{SUB,h}))$. A partial system output observed in the h-th sequence is denoted as $u_{SUB,h}(j)$.

The observed partial logical output sequences Σ_{SUB} are sequences of partial system outputs $u_{SUB}(j)$ that are introduced in Definition 22. In order to determine Σ_{SUB} for given I/O-subset SUB and Σ, the following *projection function* is defined.

Definition 30 (Projection function). Given an I/O-subset $SUB \subseteq IO$ and an observed output sequence $\sigma \in \Sigma$, the projection function $proj_{SUB}(\sigma)$ replaces each $u(j)$ in σ by the corresponding partial system output $u_{SUB}(j)$ and subsequently, $\forall u_{SUB}(j)$ with $u_{SUB}(j) = u_{SUB}(j-1)$, $u_{SUB}(j)$ is removed. The projected output sequence represents a partial output sequence σ_{SUB}.

The projection function after Definition 30 determines the logical system outputs of a subsystem with respect to the observed outputs of the entire closed-loop DES. The function is a simplification of the timed projection function, given in Definition 24, since time spans $\delta(j)$ are not considered. The resulting partial logical output sequences σ_{SUB} represents logical behavior of the considered subsystem.

Example 4.1 (Projection function). Given $IO = \{io_1, io_2, io_3\}$, the I/O-subset $SUB = \{io_1, io_2\}$, and the output sequence $\sigma \in \Sigma$ with

$$\sigma = \left(\begin{pmatrix} 0 \\ 0 \\ 0 \end{pmatrix}, \begin{pmatrix} 0 \\ 0 \\ 1 \end{pmatrix}, \begin{pmatrix} 1 \\ 1 \\ 1 \end{pmatrix}, \begin{pmatrix} 1 \\ 0 \\ 0 \end{pmatrix} \right).$$

The partial logical output sequence σ_{SUB} is determined as

$$\sigma_{SUB} = proj_{SUB}(\sigma) = \left(\begin{pmatrix} 0 \\ 0 \\ - \end{pmatrix}, \begin{pmatrix} 1 \\ 1 \\ - \end{pmatrix}, \begin{pmatrix} 1 \\ 0 \\ - \end{pmatrix} \right).$$

The comparison of partial output sequences $\sigma_{SUB,i}$ and $\sigma_{SUB,k}$ with $\sigma_{SUB,i}, \sigma_{SUB,k} \in \Sigma_{SUB}$ is made by the *equality function*. This function is defined in the following.

Definition 31 (Equality function).

$$equal(\sigma_{SUB,i}, \sigma_{SUB,k}) = \begin{cases} true & \text{if } u_{SUB,i}(j) = u_{SUB,k}(j), \forall 1 \leq j \leq \min(l_{SUB,i}, l_{SUB,k}) \\ false & \text{else} \end{cases} \tag{4.4}$$

with the partial system outputs $u_{SUB,i}(j), u_{SUB,k}(j)$ of $\sigma_{SUB,i}$ and $\sigma_{SUB,k}$, respectively, and $l_{SUB,i}, l_{SUB,k}$ are the number of system outputs contained in related partial sequences $\sigma_{SUB,i}$ and $\sigma_{SUB,k}$.

The equality function compares all partial system output $u_{SUB,i}(j)$ and $u_{SUB,k}(j)$ of two different partial sequences. If all partial system outputs are equal $u_{SUB,i}(j) = u_{SUB,k}(j)$ for all relevant j, then the sequences $\sigma_{SUB,i}$ and $\sigma_{SUB,k}$ are considered as equal. In general, the sequences can differ in length $l_{SUB,i} \neq l_{SUB,k}$. Then, the first $\min(l_{SUB,i}, l_{SUB,k})$ partial system outputs of both sequences are compared. Consequently, two sequences can be considered as equal even if they have different lengths. This is reasonable since all sequence represent the same basic system operation. Differences in length do not indicate a non-causal behavior but rather different system run times due to differing system parametrization, for instance. If $\exists j$ with $1 \leq j \leq \min(l_{SUB,i}, l_{SUB,k})$ such that $u_{SUB,i}(j) \neq u_{SUB,k}(j)$, then the sequences contain differing partial system outputs. Consequently, the sequences are considered as unequal. The equality function allows to classify the observed partial output sequences into *sequence classes*.

Definition 32 (Sequence classes). The observed partial output sequences, related to an I/O-subset *SUB* of a DES, are classified into s sequence classes

$$\Sigma_{SUB,class_1}, \Sigma_{SUB,class_2}, \ldots, \Sigma_{SUB,class_r}, \ldots, \Sigma_{SUB,class_s} \quad (4.5)$$

with the r-th sequence class given as

$$\Sigma_{SUB,class_r} = \{\sigma_{SUB,i}, \sigma_{SUB,k} \mid equal(\sigma_{SUB,i}, \sigma_{SUB,k}) = true\}. \quad (4.6)$$

A sequence class $\Sigma_{SUB,class_r}$ contains partial output sequences that are equal according to Equation 4.4. The number of sequence classes s depends on the comparison result of all partial sequences. In general, $1 \leq s \leq p$ holds with p denoting the number of observed partial output sequences contained in Σ_{SUB}.

Definition 33 (Dominant sequence class). Given an I/O-subset *SUB* and the s related sequence classes $\Sigma_{SUB,class_1}, \Sigma_{SUB,class_2}, \ldots, \Sigma_{SUB,class_s}$, the *dominant sequence class* $\Sigma_{SUB,dom}$ is determined as $\Sigma_{SUB,dom} = \Sigma_{SUB,class_i}$ such that $|\Sigma_{SUB,class_i}| \geq |\Sigma_{SUB,class_k}|$ $\forall \Sigma_{SUB,class_k} \in \{\Sigma_{SUB,class_1}, \Sigma_{SUB,class_2}, \ldots, \Sigma_{SUB,class_s}\}$ with $|\Sigma_{SUB,class_i}|$ and $|\Sigma_{SUB,class_k}|$ denoting the number of observed partial sequences contained in $\Sigma_{SUB,class_i}$ and $\Sigma_{SUB,class_k}$, respectively.

The dominant sequence class $\Sigma_{SUB,dom}$ represents the sequence class $\Sigma_{SUB,class_i}$ that contains the largest number of equal partial output sequences. It is the prevalent observed logical behavior of a considered subsystem. The comparison results are used to derive the causality degree of an I/O-subset *SUB*. Therefore, the *I/O-subset distance* is introduced by the following equation. Given an I/O-subset *SUB*, the observed partial output sequences Σ_{SUB} and the related dominant sequence class $\Sigma_{SUB,dom}$, the I/O-subset distance is determined as

$$d(SUB) = 1 - \frac{|\Sigma_{SUB,dom}|}{|\Sigma_{SUB}|} \quad (4.7)$$

with $|\Sigma_{SUB,dom}|$ and $|\Sigma_{SUB}|$ denoting the number of contained observed partial output sequences, respectively. An I/O-subset *SUB* for which $d(SUB) = 0$ holds is called *causal I/O-subset*. All observed partial sequences of the subsystem related to *SUB* are equal. The subsystem performs the same strictly logical sequential behavior in each observed system evolution. If $d(SUB) > 0$, at least one observed partial sequence exists that differs from the dominant ones. This deviation can result from non-causal behavior or noisy observations.

Example 4.2 (I/O-subset distance). Given the two I/O-subsets $SUB_A = \{io_1, io_2\}$, $SUB_B = \{io_1, io_3\}$, and the observed output sequences $\Sigma = \{\sigma_1, \sigma_2\}$ with

$$\sigma_1 = \left(\begin{pmatrix}0\\0\\0\end{pmatrix}, \begin{pmatrix}1\\0\\0\end{pmatrix}, \begin{pmatrix}1\\0\\1\end{pmatrix}, \begin{pmatrix}1\\1\\0\end{pmatrix}, \begin{pmatrix}1\\1\\1\end{pmatrix} \right), \sigma_2 = \left(\begin{pmatrix}0\\0\\0\end{pmatrix}, \begin{pmatrix}1\\0\\1\end{pmatrix}, \begin{pmatrix}1\\1\\0\end{pmatrix}, \begin{pmatrix}1\\1\\1\end{pmatrix} \right).$$

For I/O-subset SUB_A, the observed partial output sequences $\Sigma_{SUB_A} = \{\sigma_{SUB_A,1}, \sigma_{SUB_A,2}\}$ are determined as

$$\sigma_{SUB_A,1} = \left(\begin{pmatrix} 0 \\ 0 \\ - \end{pmatrix}, \begin{pmatrix} 1 \\ 0 \\ - \end{pmatrix}, \begin{pmatrix} 1 \\ 1 \\ - \end{pmatrix} \right), \sigma_{SUB_A,2} = \left(\begin{pmatrix} 0 \\ 0 \\ - \end{pmatrix}, \begin{pmatrix} 1 \\ 0 \\ - \end{pmatrix}, \begin{pmatrix} 1 \\ 1 \\ - \end{pmatrix} \right).$$

Since the partial system outputs in both partial sequences are equal, one sequence class can be determined that contains both partial sequences $\Sigma_{SUB_A,class_1} = \{\sigma_{SUB_A,1}, \sigma_{SUB_A,2}\}$. The dominant sequence class results in $\Sigma_{SUB_A,dom} = \Sigma_{SUB_A,class_1}$ and the distance metric is determined as $d(SUB_A) = 0$. For I/O-subset SUB_B, the observed partial output sequences $\Sigma_{SUB_B} = \{\sigma_{SUB_B,1}, \sigma_{SUB_B,2}\}$ are determined as

$$\sigma_{SUB_B,1} = \left(\begin{pmatrix} 0 \\ - \\ 0 \end{pmatrix}, \begin{pmatrix} 1 \\ - \\ 0 \end{pmatrix}, \begin{pmatrix} 1 \\ - \\ 1 \end{pmatrix}, \begin{pmatrix} 1 \\ - \\ 0 \end{pmatrix}, \begin{pmatrix} 1 \\ - \\ 1 \end{pmatrix} \right),$$
$$\sigma_{SUB_B,2} = \left(\begin{pmatrix} 0 \\ - \\ 0 \end{pmatrix}, \begin{pmatrix} 1 \\ - \\ 1 \end{pmatrix}, \begin{pmatrix} 1 \\ - \\ 0 \end{pmatrix}, \begin{pmatrix} 1 \\ - \\ 1 \end{pmatrix} \right).$$

The comparison after Equation 4.4 is made for the first $l_{SUB_B,2} = 4$ partial system outputs. Since $u_{SUB_B,1}(2) \neq u_{SUB_B,2}(2)$, $equal(\sigma_{SUB_A,1}, \sigma_{SUB_A,2}) = false$. The observed partial output sequences are not equal. As a result, two sequences classes $\Sigma_{SUB_B,class_1} = \{\sigma_{SUB_B,1}\}$ and $\Sigma_{SUB_B,class_2} = \{\sigma_{SUB_B,2}\}$ are determined. Since $|\Sigma_{SUB_B,class_1}| = |\Sigma_{SUB_B,class_2}|$, either $\Sigma_{SUB_B,class_1}$ or $\Sigma_{SUB_B,class_2}$ can be selected as dominant sequence class. Here, $\Sigma_{SUB_B,dom} = \Sigma_{SUB_B,class_1}$ is chosen and the distance metric is determined as $d(SUB_B) = 0.5$. Finally, the computations show that SUB_A represents a causal I/O-subset, while SUB_B does not.

In order to determine the causal I/O-subsets of a DES, the distance metric for all possible combinations of I/O-subsets needs to be determined. The number of possible I/O-subset combinations b is given as

$$b(m) = \sum_{i=1}^{m} \binom{m}{i} \tag{4.8}$$

with the number of controller I/Os m. For the BMS, with $m = 73$, in total $b(73) = 9.44 \cdot 10^{21}$ I/O-subsets need to be checked. This poses impracticable high computational efforts. In order to overcome this problem, the distance of I/O-pairs (io_i, io_j), $io_i, io_j \in IO$, is checked in the following. I/Os for which $d(io_i, io_j) = 0$ holds are considered as causally related. The idea is to determine the distance for all pairs of I/Os (io_i, io_j) and to compose causal I/O-subsets out of determined causal I/O-pairs. Therefore, the distance metric needs to be adapted and formulated in terms of an *I/O-pair distance*

$$d(io_i, io_j) = 1 - \frac{|\Sigma_{\{(io_i, io_j)\},dom}|}{|\Sigma_{\{(io_i, io_j)\}}|} \tag{4.9}$$

with $|\Sigma_{\{(io_i,io_j)\},dom}|$ and $|\Sigma_{\{(io_i,io_j)\}}|$ denoting the number of contained observed partial output sequences with respect to $SUB = \{io_i, io_j\}$, respectively. Determining the distance after Equation 4.9 for all possible I/O-pairs leads to the *I/O-pair distance matrix*

$$D = \begin{pmatrix} d(io_1, io_1) & d(io_1, io_2) & \dots & d(io_1, io_m) \\ d(io_2, io_1) & d(io_2, io_2) & \dots & d(io_2, io_m) \\ \vdots & \vdots & \ddots & \vdots \\ d(io_m, io_1) & d(io_m, io_2) & \dots & d(io_m, io_m) \end{pmatrix}. \tag{4.10}$$

The matrix D is quadratic $\mathbb{R}^{m \times m}$ and symmetric, because $d(io_i, io_j) = d(io_j, io_i)$ holds. The elements of the main diagonal are $d(io_i, io_i) = 0$. This results from the causality of an I/O to itself.

Example 4.3 (I/O-pair distance matrix). Given the I/O-subsets SUB_A, SUB_B, and the observed output sequences $\Sigma = \{\sigma_1, \sigma_2\}$ from Example 4.2, the I/O-pair distance matrix is determined as

$$D = \begin{pmatrix} 0 & 0 & 0.5 \\ 0 & 0 & 0 \\ 0.5 & 0 & 0 \end{pmatrix}.$$

A causal I/O-subset SUB_{cau} is determined by combining causally related I/O-pairs (io_i, io_j). This is done according to the following equation:

$$SUB_{cau} = \{io_i, io_j \in IO \mid d(io_i, io_j) = 0\}. \tag{4.11}$$

The set of all possible causal I/O-subsets $\forall io_i, io_j \in IO$ are denoted as CAU.

Example 4.4 (Causal I/O-subset). Given $IO = \{io_1, io_2, io_3\}$ and the distance matrix D from Example 4.3, the set of causal I/O-subsets is determined as

$$CAU = \{\{io_1\}, \{io_2\}, \{io_3\}, \{io_1, io_2\}, \{io_2, io_3\}\}.$$

In order to deal with noisy data, it is basically possible to relax Equation 4.11 such that I/Os are grouped with respect to a given distance threshold $d(io_i, io_j) \leq d_{max}$. Therefore, a parameter $d_{max} \geq 0$ needs to be selected up to which an I/O-pair (io_i, io_j) can still be considered as causal related. Due to the fact that in general only little noise is contained in observed logical output sequences and no rule exists to select d_{max}, the most conservative value $d_{max} = 0$ is applied in this thesis.

The procedure of determining causal I/O-subsets by combining causal I/O-pairs will be justified by a theorem in the following. In particular, it will be shown that if $d(io_i, io_j) = 0$, $\forall io_i, io_j \in SUB$, then $d(SUB) = 0$ holds. Therefore, the *inverse projection function* for the projection, after Definition 30, is given.

Definition 34 (Inverse projection function). Given two partial output sequences σ_{SUB_1} and σ_{SUB_2} related to the IO-subsets SUB_1 and SUB_2, the inverse projection function is defined as

$$\begin{aligned} proj^{-1}_{(SUB_1 \cup SUB_2)}(\sigma_{SUB_1}, \sigma_{SUB_2}) = \{\sigma_{(SUB_1 \cup SUB_2)} \mid & (proj_{SUB_1}(\sigma_{(SUB_1 \cup SUB_2)}) = \sigma_{SUB_1}) \wedge \\ & (proj_{SUB_2}(\sigma_{(SUB_1 \cup SUB_2)}) = \sigma_{SUB_2})\} \end{aligned} \tag{4.12}$$

with $proj_{SUB_1}(\sigma_{SUB_1 \cup SUB_2})$ and $proj_{SUB_2}(\sigma_{SUB_1 \cup SUB_2})$ representing projections according to Definition 30.

The inverse projection function determines a set of partial output sequences $\sigma_{(SUB_1 \cup SUB_2)}$ that can be projected to σ_{SUB_1} and σ_{SUB_2}, respectively. The operation can be performed for the observed sequence σ with respect to an arbitrary number of I/O-subsets

$$proj^{-1}_{(SUB_1 \cup SUB_2 \cup SUB_3 \cup \ldots)}(\sigma_{SUB_1}, \sigma_{SUB_2}, \sigma_{SUB_3}, \ldots). \tag{4.13}$$

Example 4.5 (Inverse projection function). Given the partial output sequences $\sigma_{\{(io_1,io_2)\}}$ and $\sigma_{\{(io_2,io_3)\}}$ as

$$\sigma_{\{(io_1,io_2)\}} = \left(\begin{pmatrix} 0 \\ 0 \\ - \end{pmatrix}, \begin{pmatrix} 1 \\ 0 \\ - \end{pmatrix} \right), \sigma_{\{(io_2,io_3)\}} = \left(\begin{pmatrix} - \\ 0 \\ 0 \end{pmatrix}, \begin{pmatrix} - \\ 0 \\ 1 \end{pmatrix} \right),$$

the inverse projected partial output sequences are determined as

$$proj^{-1}_{\{io_1,io_2,io_3\}}(\sigma_{\{(io_1,io_2)\}}, \sigma_{\{(io_2,io_3)\}}) = \left\{ \left(\begin{pmatrix} 0 \\ 0 \\ 0 \end{pmatrix}, \begin{pmatrix} 1 \\ 0 \\ 0 \end{pmatrix}, \begin{pmatrix} 1 \\ 0 \\ 1 \end{pmatrix} \right), \left(\begin{pmatrix} 0 \\ 0 \\ 0 \end{pmatrix}, \begin{pmatrix} 0 \\ 0 \\ 1 \end{pmatrix}, \begin{pmatrix} 1 \\ 0 \\ 1 \end{pmatrix} \right), \left(\begin{pmatrix} 0 \\ 0 \\ 0 \end{pmatrix}, \begin{pmatrix} 1 \\ 0 \\ 1 \end{pmatrix} \right) \right\}.$$

In general, the inverse projection of a number of partial output sequences does not result in a single sequence. Rather a set of possible sequences may be determined that can be again projected to the given partial output sequences. This was illustrated by Example 4.5, in which three different output sequences were determined by the inverse projection. However, in the special case that the inverse projection is performed based on the pair-wise projected sequences $\sigma_{\{(io_i,io_j)\}}$ for all I/O-pairs (io_i, io_j) that can be determined for a given I/O-subset SUB, the partial sequence σ_{SUB} can be unambiguously reconstructed. This is demonstrated by the following continuation of Example 4.5.

Example 4.5 (Cont.). Given the partial output sequences $\sigma_{\{(io_1,io_2)\}}$ and $\sigma_{\{(io_2,io_3)\}}$, as presented, and $\sigma_{\{(io_1,io_3)\}}$ as

$$\sigma_{\{(io_1,io_3)\}} = \left(\begin{pmatrix} 0 \\ - \\ 0 \end{pmatrix}, \begin{pmatrix} 1 \\ - \\ 0 \end{pmatrix}, \begin{pmatrix} 1 \\ - \\ 1 \end{pmatrix} \right),$$

the result of the inverse projection is

$$proj^{-1}_{\{io_1,io_2,io_3\}}(\sigma_{\{(io_1,io_2)\}}, \sigma_{\{(io_2,io_3)\}}, \sigma_{\{(io_1,io_3)\}}) = \left\{ \left(\begin{pmatrix} 0 \\ 0 \\ 0 \end{pmatrix}, \begin{pmatrix} 1 \\ 0 \\ 0 \end{pmatrix}, \begin{pmatrix} 1 \\ 0 \\ 1 \end{pmatrix} \right) \right\}.$$

The example shows that the logical information contained in the sequence σ_{SUB}, with $SUB = \{io_1, io_2, io_3\}$, is completely represented by the partial sequences $\sigma_{\{(io_1,io_2)\}}$, $\sigma_{\{(io_2,io_3)\}}$, and $\sigma_{\{(io_1,io_3)\}}$. The inverse projection, based on these partial sequences, allows to unambiguously reconstruct the sequence σ_{SUB}. This mapping characteristic is used in the following to show that the causality of a given I/O-subset SUB can be determined by investigating the causalities for all I/O-pairs of SUB.

Theorem 4. Given an I/O-subset SUB, if $d(io_i, io_j) = 0 \; \forall io_i, io_j \in SUB$, then $d(SUB) = 0$.

The proof is given in Appendix A. The idea behind the proof is that all partial sequences $\Sigma_{\{(io_i,io_j)\}}$ of an I/O-pair (io_i, io_j) need to be equal such that $d(io_i, io_j) = 0$ holds. Since the partial sequences $\Sigma_{\{(io_i,io_j)\}}$ for all I/O-pairs (io_i, io_j) of a given I/O-subset SUB can be unambiguously inversely projected to Σ_{SUB}, the resulting sequences Σ_{SUB} are likewise equal and the distance of SUB results to $d(SUB) = 0$. The consequence of Theorem 4 is that it is not necessary to determine the distance for each possible I/O-subset in order to find subsets SUB with $d(SUB) = 0$. Instead, it is sufficient to make the distance calculations for the I/O-pairs (io_i, io_j) and to determine the causal I/O-subsets by combining I/Os such that $d(io_i, io_j) = 0$ holds, $\forall io_i, io_j \in SUB$, according to Equation 4.11. The required calculations for determining causal related I/Os are the determination of the distance matrix D. Therefore, the distance of

$$b'(m) = \binom{m}{2}, m > 1 \tag{4.14}$$

I/O-pair combinations need to be calculated. For the BMS with $m = 73$, the number of required calculations results in $b'(73) = 2628$. Since $b'(73) \ll b(73)$, the computation efforts for distance calculations can be significantly reduced.

4.2.2 Causal Partitioning Algorithm

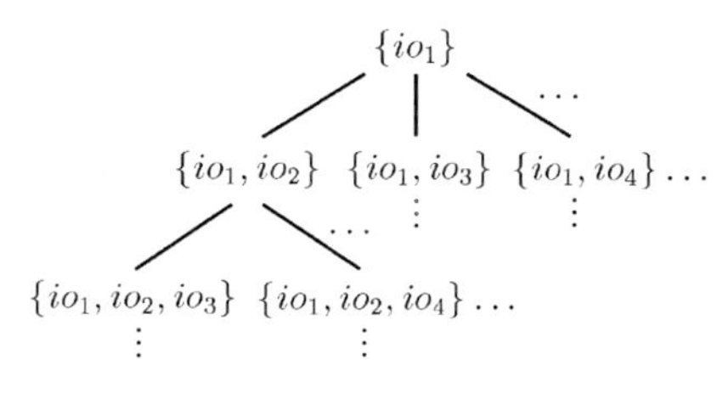

Figure 4.2: Example search tree with root node $\{io_1\}$

In the following, the causal partitioning algorithm is introduced. It implements the determination of causal I/O-subsets CAU, according to Equation 4.11. In a first step, the distance matrix D for the controller I/Os and the observed output sequences is determined. This is done for all possible I/O-pairs according to Equation 4.9. Then, the algorithm uses the I/O-pair distances to determine causal I/O-subsets $\forall io_i \in IO$. The determination of causal I/O-subsets for all I/Os represents *search problems* in search spaces that are given by tree structures, respectively. For each $io_i \in IO$, a search problem needs to be solved. An example of such a search tree is given in Figure 4.2. Each node of the tree represents an I/O-subset $SUB \subseteq IO$. The child nodes SUB' differ from the parent node SUB in exactly one I/O, respectively. The root of a search tree is given by the considered $io_i \in IO$. Out of the root, a single I/O is added to the parents node SUB on each level in order to build the child nodes. The task of the causal partitioning algorithm is to test the I/O-subsets, given by the nodes of the search tree, whether they represent a causal I/O-subset. The search effort is reduced by making use of the following proposition.

Proposition 1. If a node of the search tree, given by SUB, represents a non-causal I/O-subset, i.e. $d(SUB) > 0$, then all child nodes SUB' of SUB are non-causal I/O-subsets, i.e. $d(SUB') > 0$.

The proof is given in Appendix A. The consequence of Proposition 1 is that child nodes SUB' of non-causal I/O-subsets do not need to be tested for causality. In this case no child nodes SUB' of SUB exist such that $d(SUB') = 0$ holds. Ignoring these nodes can reduce the search efforts significantly.

Algorithm 4 Causal partitioning algorithm

Require: Controller inputs and outputs IO, observed output sequences Σ

1: $C := \emptyset$
2: Build distance matrix D for IO and Σ using Equation 4.9
3: **for all** $io_i \in IO$ **do**
4:     $SUB := \{io_i\}$
5:     **while** *true* **do**
6:         Determine $CHILDS$ by expanding SUB based on D using Algorithm 5
7:         **if** $|CHILDS| > 0$ **then**
8:             Add all $SUB' \in CHILDS$ to $STACK$
9:         **else**
10:             Add SUB to CAU
11:         **end if**
12:         **if** $|STACK| > 0$ **then**
13:             $SUB := STACK[0]$
14:         **else**
15:             Goto Line 3 for next io_i
16:         **end if**
17:     **end while**
18: **end for**
19: **return** CAU

The causal partitioning algorithm is described in Algorithm 4. It is based on the depth-first search principle. After initializing the set of causal I/O-subsets CAU, the distance matrix D is determined for the given set of controller inputs and outputs IO

and the observed output sequences Σ. Then, the algorithm starts processing the search tree for each $io_i \in IO$ in Line 3. The current I/O-subset, given by $\{io_i\}$, represents the root of the search tree. From Line 5 to Line 17, causal I/O-subsets for io_i are determined. In each iteration, the current I/O-subset SUB is expanded by Algorithm 5, in order to generate the set of child nodes $CHILD$. If child nodes can be found, they are appended to a $STACK$ in Line 8. The stack, implemented as a first-in-first-out (FIFO) buffer, contains causal I/O-subsets that may be further expanded. If no valid child nodes can be determined, then I/O-subset SUB cannot be further expanded by any I/O. Hence, SUB is a solution and added to the resulting set CAU. As long as the stack is not empty, the top I/O-subset is taken from it in Line 13 and the search procedure continues. If no nodes are left on the stack, the next root node is considered. The algorithm finish if all search trees for all root nodes $io_i \in IO$ are processed. The returned result is the set of causal I/O-subsets CAU.

Algorithm 5 expands a given I/O-subset SUB to generate causal child nodes $CHILDS$. The set of I/O-subsets $CHILDS$, which represents the child nodes, is initialized in Line 1. In the for-loop, starting in Line 2, each $io_i \in IO$ is checked whether it can be used to generate a child node SUB'. The conditions, therefore, are first, io_i is not in SUB and second, the distance of io_i to all other $io_j \in SUB$ is zero $d(io_i, io_j) = 0$. If both conditions are fulfilled, the given I/O-subset SUB is extended by io_i and the new subset is added to $CHILDS$. After processing all I/Os, the algorithm returns $CHILDS$.

Algorithm 5 Expansion algorithm

Require: Controller inputs and outputs IO, I/O-subset SUB, distance matrix D

1: $CHILDS := \emptyset$
2: **for all** $io_i \in IO$ **do**
3:    **if** $(io_i \notin SUB) \wedge ((\forall io_j \in SUB): d(io_i, io_j) = 0)$ **then**
4:       $SUB' := SUB \cup io_i$
5:       Add SUB' to $CHILDS$
6:    **end if**
7: **end for**
8: **return** $CHILDS$

In contrast to Equation 4.11, which determines all I/O-subsets with causal related I/Os, the causal partitioning algorithm returns only those I/O-subsets that are not part of any other I/O-subset, i.e. $\forall SUB_i \in C$, $\nexists SUB_j \in C$ with $SUB_i \subseteq SUB_j$. The omitted I/O-subsets are redundant since they are contained in at least one causal I/O-subset. Note that using the distance matrix to generate I/O-subsets instead of testing all I/O-subsets reduces the number of necessary comparisons but cannot avoid the worst case number of testing cases, given by b for the deep-first search. However, since child nodes of non-causal I/O-subsets are not tested and the number of causally related I/Os is in general limited, the number of actual test cases is typically significantly smaller than the number of possible ones.

Causal partitioning is especially suitable for concurrent DES that contains subsystems that perform equal operation sequence. This is often true for industrial systems, for instance manufacturing systems, in which subsystems periodically perform the same behavior in order to produce identical products. In this case, causal partitioning provides causal I/O-subsets that represent concurrent operating subsystems of the DES. The advantages of causal partitioning are the automatic and data-based operation without any required parametrization, the transparent procedure, and the acceptable computation efforts. The disadvantage of the approach originates from principle of data-based comparison of sequences. Noisy data and conditional system behavior, e.g. production recipes, can lead to differing observed output sequences for a subsystem. In this case, actually causal related I/Os of a subsystem are interpreted as non-causal, leading to *conservative* partitioning results. I/Os can exist for which only a low number of causal related I/Os can be found or even none at all. In order to deal with this limitation, the optimal partitioning approach is introduced in the following that is robust against noise and conditional behavior.

4.3 Optimal Partitioning

4.3.1 Optimization Approach

Another way to determine I/O-subsets that represent concurrent operating subsystems is to consider partitioning as an optimization problem. In the following, a new optimization based approach is presented in which the problem of determining appropriate I/O-subsets for an I/O-partition is solved by *multiple, local optimizations*. The aim is to find I/O-subsets for each I/O of a closed-loop DES that are locally optimal such that a selection of them can be synthesized to an I/O-partition with desired properties. The major advantages of local optimization are, compared to global optimization, that appropriate I/Os-subsets can be determined even if the DES is large and highly concurrent and that the computational complexity of the optimization problem is distributed among several smaller problems, which can be individually solved. In order to introduce the local optimization approach, the main ideas of global optimization have to be outlined first. In general, the aim of partitioning is to determine an I/O-partition P of a DES such that the exceeding logical behavior of a TDM is minimized while all partial automata can be logically completely identified. For the minimization of the exceeding logical behavior, modeled by a TDM, the following assumption is made:

Assumption 7 (Minimizing the exceeding logical behavior of a TDM). The exceeding logical behavior of an identified TDM is minimized if the number of I/O-subsets N, contained in the related I/O-partition $P = \{SUB_1, SUB_2, \ldots, SUB_N\}$, is minimized.

It is assumed that the exceeding logical behavior modeled by a TDM mainly depends on the number of applied partial models. This refers to the idea that with an increasing number of partial models the amount of modeled behavior, resulting from the combined behavior of all of them, also increases. Most likely, not all determined behavior combinations of the partial models refer to the original behavior of the DES. Instead, with increasing number of partial models, the exceeding modeled behavior is also enlarged.

The optimal I/O-partition, with respect to the exceeding behavior, is given with the monolithic I/O-partition, i.e. $N = 1$. It has been shown in Chapter 3 that guarantees for the precision can be given in this case. For a TDM with $N > 1$, precision decreases and no guarantees can be given any longer. Hence, N should be minimized.

In order to ensure the logical completeness of an identified TDM, the related partial automata have to be logically completely identified. It has been discussed in Chapter 3 that the logical completeness of an identified model can be estimated by means of the convergence of the related observed language. In particular, to identify a logically complete TDM, it has to be ensured that the observed logical language $L^n_{Obs,SUB}$ converges for all I/O-subsets $SUB \in P$. An optimization criterion that estimates the convergence of $L^n_{Obs,SUB}$ for all subsets of a given I/O-partition P is the *language growth criterion*. It is introduced in [Roth, 2010] as:

$$J^n(P) = \frac{1}{N} \sum_{SUB \in P} J^n(SUB) \tag{4.15}$$

with

$$J^n(SUB) = \sum_{h=2}^{p} \left(\sqrt{h} \left(|W^{n,h}_{Obs,SUB}| - |W^{n,h-1}_{Obs,SUB}| \right) \right) \tag{4.16}$$

and $W^{n,h}_{Obs,SUB}$ is the set of observed logical words for SUB with length n produced up to the h-th observed output sequence. By determining the difference $|W^{n,h}_{Obs,SUB}| - |W^{n,h-1}_{Obs,SUB}|$ between the h-th and $h-1$-th output sequence, the development of the language growth is quantified. The first output sequences $h = 1$ is not considered in order to have a relative measure for the convergence. The factor $\sqrt{h}$ is applied to punish new observations that are made closer to the observation horizon. If the value of $J^n(P)$ for an I/O-partition is comparable low, the observed partial languages $L^n_{Obs,SUB}$ are highly converging $\forall SUB \in P$.

Based on Assumption 7, the language language growth criterion given in Equation 4.15, and the requirement for the validness of I/O-partitions, the *global optimization problem* can be formulated as

$$\text{minimize } N \tag{4.17}$$

subject to

$$J^n(P) \leq J_{max}, \tag{4.18}$$

$$(\forall io \in IO)(\exists SUB \in P)\colon (io \in SUB) \tag{4.19}$$

with N as the number of I/O-subsets $SUB_1, SUB_2, \ldots, SUB_N$ contained in the I/O-partition $P = \{SUB_1, SUB_2, \ldots, SUB_N\}$, $SUB_i \subseteq IO$, and J_{max} is the selectable optimization parameter that constraints the language growth. An I/O-partition P is a solution of the global optimization problem if the number of contained I/O-subsets N is minimal while the language growth is bounded by the given threshold J_{max} and for all I/Os there exists at least one I/O-subset in P in which the I/O is contained. The parameter J_{max} is used to indirectly adjust the number of false detections that will be obtained with the resulting identified model during fault diagnosis. While the first optimization constraint follows directly from Equation 4.15, the second constraint is necessary to ensure the validness of an I/O-partition.

The presented optimization problem is rather difficult to be solved due to the large solution space that results from the high number of possible valid I/O-partitions P. It has been shown by Equation 4.1 that the size of the unconstrained solution space depends on the number of I/Os m and on the number of I/O-subsets N. For highly concurrent systems, in which N is rather large, the solutions space of the problem increases likewise. It would become even larger if shared I/Os are allowed for a solution partition P. In order to overcome these problems, a local optimization approach is proposed in the following. The approach distributes the optimization complexity among several, smaller problems, which can be solved in parallel. In that way, I/O-partitions with shared I/Os can be determined for large and highly concurrent DES. As a first step, the local optimization aim must be formulated in such a way that N will be minimized on the global level. This aim relies on the following assumption:

Assumption 8 (Minimization of N). The number of I/O-subsets N contained in a valid I/O-partition $P = \{SUB_1, SUB_2, \ldots, SUB_N\}$ is minimized if $(\forall io \in IO)(\exists SUB \in P)$ such that $io \in SUB$ and $|SUB|$ is maximal.

It is assumed that an I/O-partition N can be minimized if for all I/Os of the closed-loop DES a subset SUB in P exists that contains a maximum number of I/Os, including the considered one. In general, several I/O-subsets may exist that all contain a given I/O. By adding the largest one of them to an I/O-partition P, it becomes most likely that all I/Os can be represented by a minimum number of subsets N. The more I/Os are covered by the determined I/O-subset, the less I/O-subsets are required for a valid partition. The consequence for local optimization is that subsets need to be determined for each I/O such that the number of contained I/Os is maximal. The optimal solution, when no constraints are considered, is given with the monolithic I/O-subset $SUB = IO$. In this case, $|SUB|$ is maximum for all I/Os and $N = 1$ results for the valid I/O-partition $P = \{IO\}$.

In addition to the precision aim, the constraint for the language growth needs to be adapted for the local optimization, accordingly. Substituting Equation 4.15 into Equation 4.18 leads to

$$J^n(SUB_1) + J^n(SUB_2) + \ldots + J^n(SUB_N) \leq N \cdot J_{max} \tag{4.20}$$

which holds, if

$$\begin{aligned} J^n(SUB_1) &\leq J_{max}, \\ J^n(SUB_2) &\leq J_{max}, \\ &\vdots \\ J^n(SUB_N) &\leq J_{max}. \end{aligned} \tag{4.21}$$

If the language growth for each $SUB_i \in P$ is bounded such that $J^n(SUB_i) \leq J_{max}$ holds, the language growth of a resulting I/O-partition P is bounded by $N \cdot J_{max}$. In this way, independently determined I/O-subsets can be used to compose an I/O-partition for a given J_{max}. In general, there may exist I/O-partitions P with minimum N and $J^n(P) \leq J_{max}$ but $J^n(SUB_i) \leq J_{max}$ does not hold $\forall SUB_i \in P$. These solution partitions cannot be determined by means of the presented procedure.

In accordance with the Assumption 8 and the inequalities given in Equation 4.21, the *local optimization problem* can be formulated as:

$$\text{maximize } |SUB| \quad (4.22)$$

subject to

$$J^n(SUB) \leq J_{max} \quad (4.23)$$

with $SUB \subseteq IO$ and the optimization parameter J_{max}, which constraints the language growth. An I/O-subset SUB is a solution of the local optimization problem if the number of contained I/Os is maximal while the language growth is below the given threshold J_{max}. In that way, I/O-subsets are determined that can be used to build an I/O-partition which is complete and highly precise.

The optimization problem of Equation 4.22 is solved for each I/O of the closed-loop DES, respectively. This ensures that optimal I/O-subsets are determined for all of them. The optimization problem represents a *Knapsack problem*.[11] The task is to put a number of items (I/Os) into a knapsack (I/O-subset) such that its value (size) is maximized while its weight (language growth) is bounded. Terms in brackets are subject to the given I/O-subset determination problem. In the context of partitioning, each I/O has the same value. The value of the I/O-subset corresponds to the number of I/Os it contains, i.e. its size. This value is to be maximized.

The Knapsack problem is known to be NP-complete [Kellerer et al., 2004]. For a given I/O, in total

$$b''(m) = 1 + \sum_{i=1}^{m-1} \binom{m-1}{i} \quad (4.24)$$

optimal I/O-subset combinations are possible. For the BMS with $m = 73$, the number of possible combination for each I/O results in $b''(73) = 4.72 \cdot 10^{21}$. However, the constrained search space is significant smaller with respect to the choice of J_{max}. In order to speed up the optimization procedure, causal partitions can be used as initial solutions to the optimization problems instead of running it for single I/Os. Nevertheless, the solution space is still too large to simply check all possibilities by a brute force algorithm. Hence, an appropriate optimization algorithm that solves the given Knapsack problem needs to be applied.

4.3.2 Optimal Partitioning Algorithm

The algorithm used to solve the introduced Knapsack problem is a *stochastic hill climbing algorithm*. Hill climbing algorithms are classified as local search algorithm [Russell and Norvig, 2010]. They are inspired by the problem to climb up a hill, step-by-step. In order to find the optimal solution, i.e. the top of a hill, the algorithm takes exclusively steps such that a next determined position is an improvement of the current one. This is done as long as a current position can be improved. A general problem of local search

[11] See [Kellerer et al., 2004] for a comprehensive overview about the Knapsack problem.

algorithms is that they may get stuck in local optimal solutions. In order to overcome this problem, the algorithm, presented in the following, has a stochastic component. The selection of next positions is made randomly instead of a deterministic selection, as it is done in classical search algorithms. This allows to cover a wider range of the search space and increases the chance to find the global optimum.

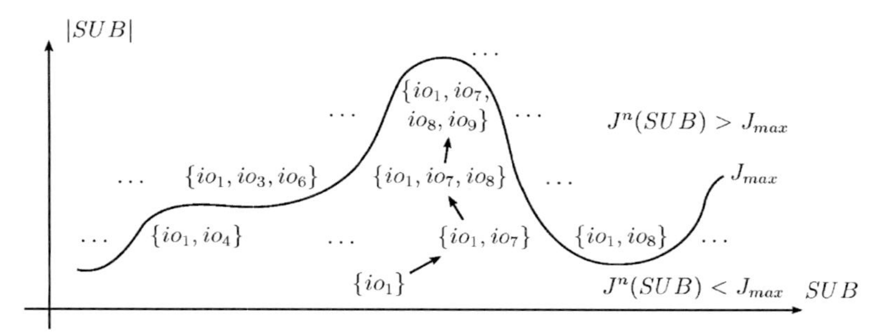

Figure 4.3: Example of a stochastic hill climbing run for $SUB_{init} = \{io_1\}$

With respect to the optimization problem, given by the Equations 4.22 and 4.23, a position is represented by the current I/O-subset SUB_{cur}. The 'height' of a position corresponds to the value of the target function $|SUB_{cur}|$ that is to be maximized. The shape of the actual search space is given by the optimization parameter J_{max}. This value limits the overall search space and delimits feasible solutions with $J^n(SUB_{cur}) \leq J_{max}$ from infeasible solutions with $J^n(SUB_{cur}) > J_{max}$. The aim of the hill climbing algorithm is to find an optimal position SUB_{cur} for a given initial position SUB_{init} such that $|SUB_{cur}|$ is maximum and $J^n(SUB_{cur}) \leq J_{max}$ holds. In Figure 4.3, an example for a stochastic hill climbing run is depicted. The initial position SUB_{init} is given by io_1. In each iteration of the algorithm, an $io \in IO \setminus SUB_{cur}$ is added to the current position SUB_{cur}, respectively, to generate the set of feasible next positions $NEXT$. For all feasible next position $SUB_{next} \in NEXT$, $J^n(SUB_{next}) \leq J_{max}$ holds. Then, one of the next positions is randomly selected and set as current position. The algorithm extends SUB_{cur} as long as feasible next positions can be determined. In the figure, the algorithm stops after reaching $SUB_{cur} = \{io_1, io_7, io_8, io_9\}$. This position cannot be further improved by the algorithm without violating the optimization constrained given by J_{max}. The I/O-subset SUB_{cur} represents to optimal solution for this run. The optimization of SUB_{init} can be executed several times in order to find different optimal solutions. This increases the probability to find the global optimal solution. The optimal partitioning algorithm proposed in the following is a modification after the hill climbing algorithm proposed in [Russell and Norvig, 2010]. It determined optimal I/O-subsets for a given initial I/O-subset SUB_{init} and adds them to the set of optimal I/O-subsets OPT.

The stochastic hill climbing algorithm requires the initial position $SUB_{init} \subseteq IO$, the controller inputs and outputs $IO = \{io_1, io_2, \ldots, io_m\}$, the optimization parameter J_{max},

Algorithm 6 Stochastic hill climbing algorithm

Require: Initial position SUB_{init}, set of controller inputs and outputs IO, optimization parameter J_{max}, language parameter n, observed output sequences Σ

1: **while** $stopCrit = false$ **do**
2: $SUB_{cur} := SUB_{init}$
3: Spot feasible next positions $NEXT$ using Algorithm 7 based on SUB_{cur}, IO, J_{max}, n, and Σ
4: **if** $|NEXT| > 0$ **then**
5: $SUB_{cur} := random(NEXT)$
6: Goto Line 3 and continue the walk
7: **end if**
8: Add SUB_{cur} to OPT
9: Update $stopCrit$
10: **end while**
11: **return** OPT

language parameter n, and the observed output sequences $\Sigma = \{\sigma_1, \sigma_2, \dots \sigma_p\}$. The initial position is either a single I/O or a given I/O-subset. The stochastic hill climbings are performed for the given initial position SUB_{init}, as long as the stop criterion *stopCrit* is false. A hill climb starts in Line 2, by setting the current position SUB_{cur} as the given initial position SUB_{init}. Then, the feasible next positions $NEXT$ are spotted by executing Algorithm 7 for the current position SUB_{cur}. If at least one feasible next position can be determined, a random one is selected in Line 5 from $NEXT$. This ensures that in multiple optimization runs different routes are taken to avoid getting stuck in a given local optimum. After the selection, the position spotting from Line 3 is repeated. This is done as long as next positions $NEXT$ can be determined. If $|NEXT| = 0$, the climbing is finished and SUB_{cur} represents the optimal solution that is added to the output set OPT in Line 8. In Line 9, the stop criterion of the algorithm is updated. The criterion states how many times hill climbing is performed for a given initial position SUB_{init}. Possible conditions are a maximum number of runs with no generated new optimal I/O-subsets and a maximum number of runs in total. After finishing the optimization, the algorithm returns the updated set of optimal I/O-subsets OPT.

The spotting of feasible next positions is implemented by Algorithm 7. Initially, the set of feasible next positions $NEXT$ is initialized. In the for-loop, starting in Line 2, the next positions are spotted. Therefore, the given current position SUB_{cur} is extended by one I/O, respectively, that is not part of the SUB_{cur} yet. For this next position SUB_{next}, the language growth $J^n(SUB_{next})$, based on Σ, is then determined in Line 4. If $J^n(SUB_{next})$ is equal or below the given optimization parameter J_{max}, SUB_{next} is added to $NEXT$. Otherwise, SUB_{next} is discarded. After processing the I/O-subsets for all possible I/Os, the algorithm returns the set of feasible next positions $NEXT$. In case that either the monolith is reached such that $IO \setminus SUB_{cur} = \emptyset$ holds, or the top of a hill is reached $J^n(SUB_{next}) > J_{max}\ \forall SUB_{next}$, then $NEXT = \emptyset$.

The advantage of using a local search algorithm as stochastic hill climbing is that the

Algorithm 7 Spot feasible next positions

Require: Current position SUB_{cur}, controller inputs and outputs IO, optimization parameter J_{max}, language parameter n, observed output sequences Σ

1: $NEXT := \emptyset$
2: **for all** $io \in (IO \setminus SUB_{cur})$ **do**
3: $\quad SUB_{next} = SUB_{cur} \cup io$
4: $\quad$ Determine $J^n(SUB_{next})$ for Σ according to Equation 4.16
5: $\quad$ **if** $J^n(SUB_{next}) \leq J_{max}$ **then**
6: $\quad\quad$ Add SUB_{next} to $NEXT$
7: $\quad$ **end if**
8: **end for**
9: **return** $NEXT$

algorithm always finds local optimal solutions for a given initial I/O-subset. By executing the optimization algorithm several times, a wide area of the search space can be covered. However, no guarantee can be given that the global optimal solution can be found. The number of optimal solutions strongly depends on the choice of the optimization parameter J_{max}. By selecting a small J_{max}, the feasible solution space is comparable smaller, hence less optimal solutions exists. It is more likely in this case to find the global optimal solution for a given number of optimization runs. This is particular convenient since J_{max} is typically selected by a small value in order to obtain I/O-subsets with partial observed languages that are highly converging. The search space can further be reduced by taking causal I/O-subsets CAU as initial positions for the optimization. Causal I/O-subsets can be determined by the approach presented in the preceding section.

For the language parameter, $n > 1$ holds. The lower bound for n is motivated by the identification algorithm introduced in Chapter 3. In order to identify logically complete models, a $k \geq 1$ has to be determined such that the observed partial logical language $L^n_{Obs,SUB}$ of length $n = k + 1$ converges. Hence, at least $L^{n=2}_{Obs,SUB}$ needs to converge. For the upper bound of n, in general no limitation exists. Depending on the choice of n, the optimization parameter J_{max} can be adapted accordingly. However, it is reasonable to select n such that $n = k+1$ holds. In this case, the I/O-subsets are explicitly determined with respect to the costs of the observed language that is used for identification.

The presented optimal partition approach determines optimal I/O-subsets individually for each I/O of the closed-loop DES. This allows to distribute the optimization computations on an arbitrary number of computation units which significantly reduces the necessary computation times. In order to synthesize an I/O-partition out of the determined I/O-subsets, the partition synthesis approach is presented in the next section.

4.4 Partition Synthesis

Causal and optimal partitioning determine I/O-subsets CAU and OPT that represent concurrent operation subsystems of a DES. The subsets resulting from the union over CAU

and OPT are called *feasible I/O-subsets* $F = CAU \cup OPT$. In order to determine a valid I/O-partition P of the DES, a number of I/O-subsets need to be selected from F. This is the *partition synthesis*. The task of partition synthesis is to select those I/O-subsets $P \subseteq F$ such that an identified model based on the resulting I/O-partition P has minimal exceeding and non-reproducible behavior. The minimization of the non-reproducible behavior is ensured by causal and optimal partitioning. Both approaches generate I/O-subsets for which the partial logical observed languages are converging. Causal partitioning implements this by grouping only I/Os that perform the same sequential behavior such that the observed language is highly converging. Optimal partitioning explicitly bounds the language growth by a given threshold J_{max}. For minimizing the exceeding logical behavior of a model, based on P, the number of I/O-subsets N of the solution I/O-partition P is minimized, according to Assumption 7. In general, for each $io \in IO$ more than one I/O-subset SUB can be determined by the presented partitioning approaches such that $io \in SUB$ holds. Partition synthesis needs to select those I/O-subsets from F such that N of the resulting I/O-partition is minimal and P is valid. In summary, partition synthesis can be formulated as the following minimization problem:

$$\text{minimize } N \tag{4.25}$$

subject to

$$(\forall io \in IO)(\exists SUB \in P) \colon (io \in SUB) \tag{4.26}$$

with N as the number of selected I/O-subsets from F and $P \subseteq F$. In words, the aim of partition synthesis is to generate an I/O-partition $P = \{SUB_i, SUB_2, \ldots, SUB_N\}$ with $SUB_i \in F$ such that the number of selected I/O-subsets N is minimal and all I/Os of the closed-loop DES $io \in IO$ are represented by at least one I/O-subset $SUB \in P$. This minimization problem represents a *Set Cover problem*, see [Korte and Vygen, 2012]. Given a set of elements IO and a collection of sets F, the Set Cover problem is to select the smallest number of subsets from F such that all elements of IO are covered. The Set Cover problem is known to be NP-hard [Korte and Vygen, 2012]. In order to solve the Set Cover problem, a *Greedy algorithm*, proposed in [Korte and Vygen, 2012], is applied. It allows to find a best effort solution while the optimization problem need not to be explicitly solved. The Greedy algorithm is based on the following principle. In each iteration, an I/O-subset $SUB \in F$ is selected that covers the largest number of I/Os that are not yet covered by P. This procedure is continued until an I/O-partition P is found that covers all I/Os of the DES. The resulting I/O-subsets contained in P can overlap due to shared I/Os. This is a major benefit of the presented partitioning approach since shared I/Os are beneficial to minimize the exceeding logical behavior of a DES model, as shown in Chapter 3. In the following, the applied partition synthesis algorithm is presented. It is a modification of the algorithm proposed in [Korte and Vygen, 2012].

In the initial step of the algorithm, the solution I/O-partition P is initialized. Beginning from Line 2, I/O-subsets are added to P as long as not all I/Os are covered by P. The set $COVER$ represents the I/Os that are covered by the I/O-subsets contained in P. In each iteration, the I/O-subset SUB is determined such that $|SUB \setminus COVER|$ is maximum. If more than one SUB can be determined, then a random one is selected.

Algorithm 8 Partition synthesis algorithm

Require: Feasible I/O-subsets F, controller inputs and outputs IO

1: $P := \emptyset$
2: **while** $COVER \neq IO$ **do**
3: Select $SUB \in (F \setminus P)$ such that $|SUB \setminus COVER|$ is maximum
4: Add SUB to P
5: $COVER := COVER \cup SUB$
6: **end while**
7: **return** P

The selected I/O-subset SUB represents the most I/Os not yet covered by any other I/O-subset already contained in P. In the next step, SUB is added to P and $COVER$ is updated. If all I/Os are covered by the selected I/O-subsets, then $COVER = IO$ and the algorithm returns the I/O-partition P.

Example 4.6 (Partition synthesis). Given the set of controller inputs and outputs $IO = \{io_1, io_2, io_3, io_4, io_5\}$ and the set of feasible I/O-subsets $F = \{SUB_1, SUB_2, SUB_3\}$ with $SUB_1 = \{io_1, io_2\}$, $SUB_2 = \{io_2, io_3\}$ and $SUB_3 = \{io_3, io_4, io_5\}$. In the first step, SUB_3 is selected by the synthesis algorithm since $COVER = \emptyset$ and $|SUB_3 \setminus COVER| = 3$ is maximum. All other I/O-subsets cover less I/Os. The set $COVER$ is updated to $COVER = \{io_3, io_4, io_5\}$. Then, SUB_1 is added to P since $|SUB_1 \setminus COVER| = 2$ and $|SUB_2 \setminus COVER| = 1$. After this step, all I/Os are covered by the partition $COVER = IO$, hence the solution partition $P = \{SUB_3, SUB_1\}$ is found.

In Example 4.6, the Greedy algorithm found the global optimal solution of the given synthesis problem. However, since the algorithm is based on a heuristic it cannot be guaranteed in general that the global optimum can always be found. The algorithm provides one or more best effort solutions that solve the partitioning problem. In various practical applications, the algorithm led to appropriate synthesis results. This will be exemplary shown by means of the application to the BMS.

4.5 Partitioning of BMS Models

4.5.1 Causal Partitioning

The causal partitioning approach presented in Section 4.2 was applied to the BMS. The number of considered controller inputs and outputs was given as $|IO| = 73$. For automatic partitioning, $p = 50$ observed output sequences were available. Initially, the distance matrix D with 73×73 elements was determined. The computation took 99 seconds with a desktop PC.[12] The resulting matrix is illustrated in Figure 4.4. Elements of the D with $d(io_i, io_j) = 0$ are drawn white and elements with $d(io_i, io_j) > 0$ are drawn black. The white elements represent the causal I/O-pairs. Executing Algorithm 4 led to 40 causal

[12] Intel(R) Core(TM) i3-530 2.94 GHz.

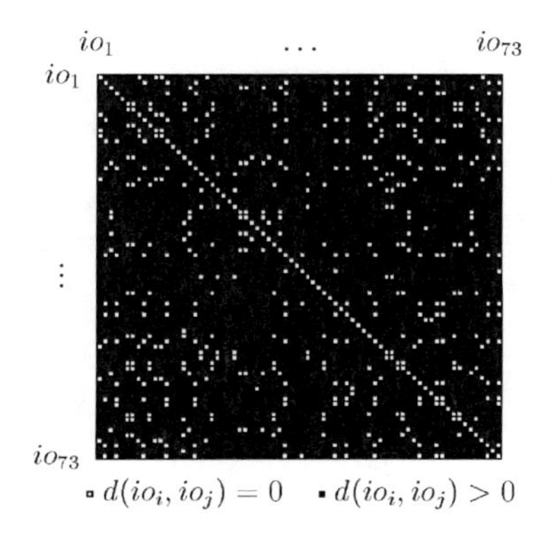

Figure 4.4: Distance matrix D of the BMS

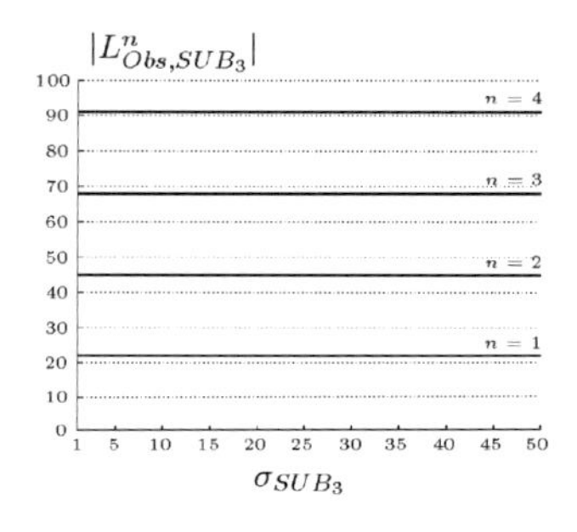

Figure 4.5: Observed language growth for the causal I/O-subset $SUB_3 \in CAU$

I/O-subsets for the BMS based on the distance matrix D. The determination took 32 minutes. In Figure 4.5, the growth of the observed partial logical language L^n_{Obs,SUB_3} for the causal I/O-subsets $SUB_3 \in CAU$ is depicted. The considered I/O-subset SUB_3 contains 11 causally related I/Os. It can be seen that the observed partial languages for lengths $1 \leq n \leq 4$ are converging after the first observed sequence. This means that after the first sequence no new words of length n were observed. Hence, the original logical language referring to SUB_3 can be considered as completely observed with respect to Σ. The high convergence of observed partial logical language is typical for causal I/O-subsets. This results from the requirement that their behavior has to be equal in all observed sequences. The illustrated language growth is representative for all causal I/O-subsets CAU determined by causal partitioning.

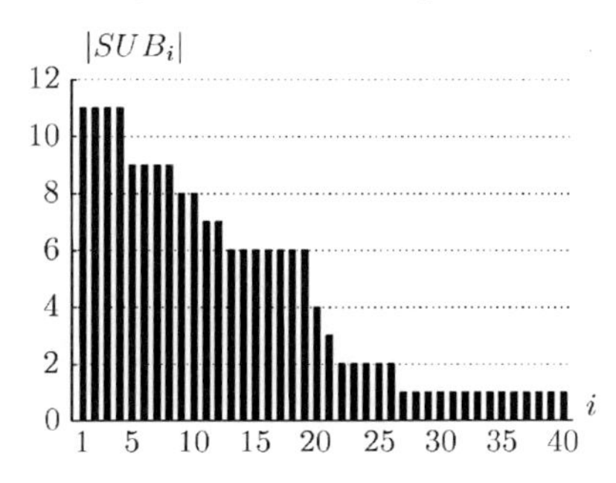

Figure 4.6: Number I/Os contained in each causal I/O-subset $SUB_i \in CAU$

In Figure 4.6, the number of I/Os $|SUB_i|$ represented by each causal I/O-subset $SUB_i \in CAU$ is illustrated. The number differs from 1 to 11 for all causal I/O-subsets.

For most of the I/Os at least one other I/O could be found such that a causal I/O-subset can be generated. However, there are 14 causal I/O-subsets that represent only one single I/O, i.e. $|SUB_i| = 1$. For these I/Os, no causal relationship to any other I/O could be determined, hence they could not be causally grouped with any other I/O. For partition synthesis, large I/O-subsets with many I/Os are beneficial to generate I/O-partitions with a low number of I/O-subsets. Therefore, optimal partitioning was applied to search for more appropriate I/O-subsets, especially for I/Os for which no causalities could be determined.

4.5.2 Optimal Partitioning

Optimal partition was based on the same output sequences Σ that were used for causal partitioning. In order to reduce the computation efforts for solving the optimization problem given in Equations 4.22 and 4.23, the 40 determined causal I/O-subsets CAU were used as initial positions for the stochastic hill climbing. This allowed to determine optimal I/O-subsets, especially for I/Os for which no or only a low degree of causality could be determined. Consequently, 40 instances of stochastic hill climbings, according to Algorithm 6, needed to be executed. The climbings are independent of each other and could be executed in parallel on different computation units. If causal I/O-subsets were not available or not used, optimal partitioning could be performed for the 73 I/Os of the BMS. Hill climbing for a given initial position was stopped if 200 hill climbing runs were performed or if after 10 successive runs no further optimal solutions could be determined. These conditions denote the stop criterion, represented by $stopCrit$ in Algorithm 6.

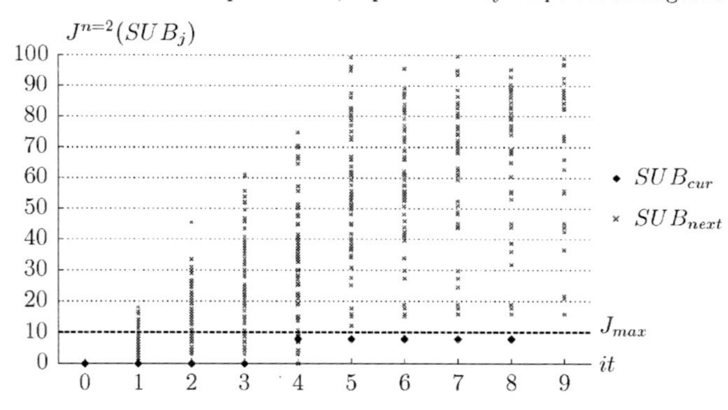

Figure 4.7: Example stochastic hill climbing run for $SUB_{init} = \{O50.10\}$ and $J_{max} = 10$

An example hill climbing run for the initial position $O50.10$, the optimization parameter $J_{max} = 10$, and the language parameter $n = 2$ is depicted in Figure 4.7. The I/Os of the BMS are labeled either by I, denoting an controller input, or by O, denoting an controller output, followed by an individual number that is uniquely defined for each I/O. In Table 4.1, the current positions SUB_{cur} and values of the cost function $J^{n=2}(SUB_{cur})$

it	SUB_{cur}	$J^{n=2}(SUB_{cur})$
0	$O50.10$	0.0
1	$O50.10$, $I1.5$	0.0
2	$O50.10$, $I1.5$, $I1.2$	0.0
3	$O50.10$, $I1.5$, $I1.2$, $O50.13$	0.0
4	$O50.10$, $I1.5$, $I1.2$, $O50.13$, $O50.12$	7.94
5	$O50.10$, $I1.5$, $I1.2$, $O50.13$, $O50.12$, $I1.4$	7.94
6	$O50.10$, $I1.5$, $I1.2$, $O50.13$, $O50.12$, $I1.4$, $O50.9$	7.94
7	$O50.10$, $I1.5$, $I1.2$, $O50.13$, $O50.12$, $I1.4$, $O50.9$, $O50.11$	7.94
8	$O50.10$, $I1.5$, $I1.2$, $O50.13$, $O50.12$, $I1.4$, $O50.9$, $O50.11$, $O50.8$	7.94

Table 4.1: Example stochastic hill climbing run for $SUB_{init} = \{O50.10\}$ and $J_{max} = 10$

J_{max}	0	5	10
$\|OPT\|$	448	1452	2208

Table 4.2: Stochastic hill climbing results for the BMS

for each hill climbing iteration it are given. The hill climb started with the given initial position $SUB_{cur} = \{O50.10\}$. $O50.10$ is one of the I/Os for which no causal relationship to any other I/O could be determined by causal partitioning. During the execution of the stochastic hill climbing algorithm, the feasible and infeasible next positions were determined and evaluated by means of $J^{n=2}(SUB_{next})$. Both, the feasible and infeasible next positions SUB_{next}, determined in Line 3 of Algorithm 7, are marked in Figure 4.7 by gray crosses. Out of the feasible ones, with $J^{n=2}(SUB_{next}) \leq J_{max}$, the next current position SUB_{cur} was randomly selected. In the example, $I1.5$ was selected after the first iteration $it = 1$, hence $SUB_{cur} = \{O50.10, I1.5\}$. The determined current positions are marked in the figure by black diamonds. One can see that a selected position is not necessarily the one with the lowest cost. For instance, at iteration $it = 4$, SUB_{cur} with $J^{n=2}(SUB_{cur}) = 7.94$ was selected, although other feasible next positions were determined with lower costs. This results from the random selection in order to avoid getting stuck in local optima and to increases the chance to find the global optimal solution. In the 9-th optimization iteration, no feasible next position $NEXT$ could be generated with $J^{n=2}(SUB_{next}) \leq J_{max}$. Hence, the hill climbing was finished and the determined optimal solution was added to OPT. After that, the optimization could be reinitialized for $SUB_{init} = \{O50.10\}$.

In Table 4.2, the results of optimal partitioning for the given parameters are summarized. Given the 40 causal I/O-subsets, 448 optimal I/Os-subsets could be determined for $J_{max} = 0$, 1452 optimal I/Os-subsets for $J_{max} = 5$ and 2208 optimal I/Os-subsets for $J_{max} = 10$. Obviously, the number of optimal I/O-subsets grows while increasing J_{max}. An increase of the optimization parameter J_{max} enlarges the feasible search space of the optimization after Equation 4.22 leading to more possible solutions. This is beneficial to find more and possible larger I/O-subsets, but also increases the chance of non-reproducible behavior and the computational efforts. One of the 40 optimizations for $J_{max} = 5$ took in average 5 hours and 44 minutes to be computed.[13] The minimum time was 21 minutes and the maximum time 19 hours and 19 minutes. The different computation times refer to the different initial conditions and the related individual size of the search space. The number of optimal I/O-subsets found for a given initial position was 36 on average, the minimum was 1 and the maximum was 133.

4.5.3 Partition Synthesis

Partition synthesis uses the feasible I/O-subsets $F = CAU \cup OPT$, determined by causal and optimal partition, to generate the I/O-partition P of the BMS. For the evaluation of the automatically determined I/O-partitions, two reference solutions, determined by experts, are available. The proposed automatic partitioning approaches do not make use of this expert information. They are exclusively used for evaluation purposes. The first expert I/O-partition available is the monolithic I/O-partition $P_{mon} = \{IO\}$. Given $n = 2$, the costs of the monolithic partition was determined as $J^{n=2}(P_{mon}) = 202413.44$. The second given partition is the distributed expert partition P_{dis} with $N = 10$ and $J^{n=2}(P_{dis}) = 7.44$ for $n = 2$. The distributed expert partition P_{dis} was determined in accordance with the modular BMS structure, shown in Figure 1.3. The modules of the BMS were considered as the subsystems. For each of them, an I/O-subset $SUB \in P_{dis}$ was created containing the related I/Os. The two exceptions are the sensor array and the transport system of the second station that are represented by one I/O-subset and the storage of fourth station that has no sensors and actuators attached. The distributed expert partition P_{dis} contains no shared I/Os.

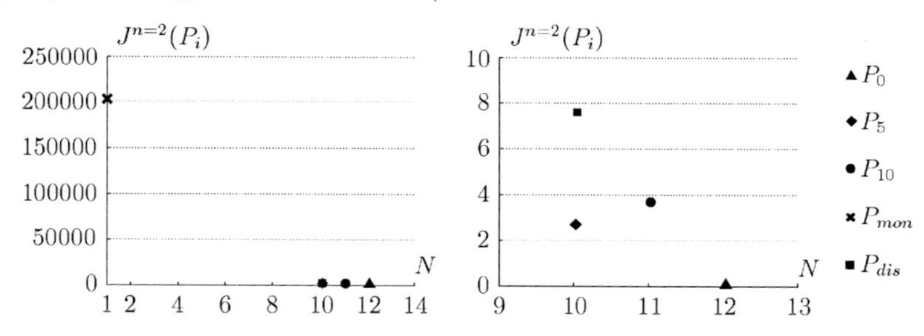

Figure 4.8: Evaluation of the automatically determined and expert I/O-partitions

The results of partition synthesis are illustrated in Figure 4.8. The evaluated I/O-partitions are the automatically determined ones P_0, P_5, and P_{10} for $J_{max} = 0$, $J_{max} = 5$, and $J_{max} = 10$, respectively, and the expert ones P_{mon} and P_{dis}. The I/O-partitions are compared with respect to their size N and the language growth criterion $J^{n=2}(P_i)$ for $n = 2$. The left diagram in the figure shows the difference between the monolithic I/O-partition $N = 1$ and the distributed I/O-partitions with $10 \leq N \leq 12$. It can be

[13] Intel(R) Core(TM) i3-530 2.94 GHz.

j	1	2	3	4	5	6	7	8	9	10
$J^{n=2}(SUB_j)$	4.00	3.73	2.83	0.0	0.0	2.65	2.45	2.93	3.16	4.58

Table 4.3: Cost of the I/O-subsets $SUB_j \in P_5$

recognized that the costs of the monolithic I/O-partition $J^{n=2}(P_{mon}) = 202413.44$ are comparable high compared to the costs of the distributed I/O-partitions. This indicates a low convergence of the observed logical language which leads to a likely incomplete model. If this incompleteness cannot be tolerated, a distributed I/O-partition needs to be selected for modeling. Note that the 'star-symbol' in the left diagram results from the overlay of the symbols for P_5 and P_{dis}. In the right diagram, the section with the distributed I/O-partitions is zoomed in. This area is marked by a dotted rectangle in the left diagram of the figure. Obviously, the automatically generated I/O-partitions do all meet their preselected J_{max}. For $N = 10$, P_5 outperforms the expert solution P_{dis} since the costs $J^{n=2}(P_5) = 2.63$ of P_5 are lower than for P_{dis} where $J^{n=2}(P_{dis}) = 7.44$.

In Table 4.3, the costs of the synthesized I/O-subsets $SUB_j \in P_5$ are shown. In general, the applied optimization parameter J_{max} represents an upper bound for the determined convergence criterion $J^{n=2}(SUB_j)$. A synthesized I/O-partition can contain I/O-subsets that outperform this requirement, e.g. SUB_5 where $J^{n=2}(SUB_5) = 0.0 \leq 5$ holds. The language growth diagrams for all I/O-subsets $SUB_j \in P_5$ were depicted in Figure 3.25, in the application section of Chapter 3. In Figure 3.24, the diagram for the related monolithic I/O-partition P_{mon} was shown. Due to the usage of the introduced Greedy algorithm, no notable computation time was required for the synthesis of P_0, P_5, and P_{10}. The optimization parameter J_{max} was chosen such that I/O-partitions are generated that are comparable with the distributed expert solution P_{dis}.

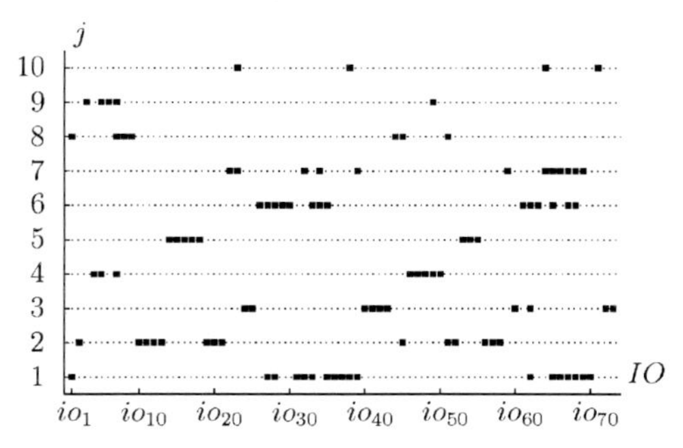

Figure 4.9: I/O-mapping of $SUB_j \in P_5$

A important advantage of the presented partitioning approach compared to other approaches is that I/O-partitions with shared I/Os can be determined for large and concurrent systems in reasonable time. An illustration of the shared I/Os of P_5 is given in Figure 4.9. A black square indicates that an $io_i \in IO$ is contained in a given I/O-subset

$SUB_j \in P_5$. In total, 52.5% of BMS I/Os are contained in 1 I/O-subset, 32.3% are contained in 2 I/O-subsets, and 15.2% are contained in 3 I/O-subsets. The high number of shared I/Os connects the behavior of the partial models and helps to minimize the modeled exceeding logical behavior of the resulting TDM.

Chapter 5

Fault Detection and Isolation using Timed Distributed DES Models

5.1 Preliminaries

A framework for *automatic model-based fault diagnosis* of DES is presented in the following. The related challenges and requirements have been discussed in Chapter 2. In summary, a fault diagnosis framework must perform online detection and isolation of faults based on logical and time fault symptoms. The required online observation of the DES needs to be passive, i.e. it must not affect the operation of the system. Furthermore, it must be possible to economically implement a diagnosis framework independent of system size and complexity. Since the costs of a model-based diagnosis system mainly depend on the construction of a proper fault diagnosis model, automatic modeling approaches, as the ones presented in Chapter 3 and 4, contribute to this aim.

Model-based fault diagnosis of DES is a well-known field of research where many contributions have been published in the last years. One of them is the diagnoser approach [Sampath et al., 1995], in which an observer automaton, called diagnoser, is used for fault detection and isolation of predefined faults. Another approach is given in [Dotoli et al., 2009]. Fault diagnosis is formulated there as an integer linear programming problem based on a special Petri net. These centralized FDI concepts, using monolithic models, have been extended to perform decentralized diagnosis [Wang et al., 2007] and distributed fault diagnosis [Genc and Lafortune, 2003]. In contrast to the so far named concepts, which explicitly model faults, [Reiter, 1987] and [Roth, 2010] proposed FDI approaches that rely on fault-free behavior models. Fault diagnosis considering timed behavior is pursued in [Hashtrudi Zad et al., 2005], [Zaytoon and Sayed-Mouchaweh, 2012], [Jiroveanu and Boel, 2006], and [Supavatanakul et al., 2006]. Modeled time behavior allows to detect and to isolate faults by means of time fault symptoms. However, no approach exists that performs timed FDI using automatically determined fault-free models with respect to the time fault symptoms early, late, and deadlock behavior. See Chapter 6 for a comprehensive literature discussion.

The model-based FDI approach proposed in this thesis for the automatic timed de-

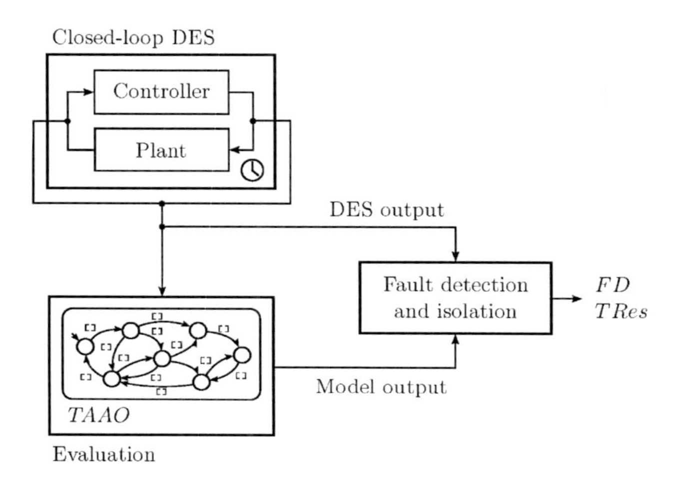

Figure 5.1: Overview model-based FDI using a monolithic *TAAO*

tection and isolation of faults is depicted in Figure 5.1. FDI is performed online during operation of the system with respect to the passively monitored DES output. The aim is to recognize and interpret fault symptoms generated by the DES in order to detect faults and isolate the fault related hardware components. Therefore, the DES output is compared with the evaluated output of the timed fault-free automaton model. The evaluator estimates the current model state and provides the time information that is required for timed FDI. Based on this, the fault state of the DES is determined, denoted as FD in the figure. In case of a detected fault, DES specific residuals are computed in order to isolate the fault. The concept of residuals is adapted in this work from fault diagnosis theory of continuous systems. They explain discrepancies between process and model to isolate faulty hardware components [Isermann, 2006]. In [Roth, 2010], logical residuals have been developed that focus on the logical fault symptoms unexpected and missed behavior. They are discussed in Chapter 6. In this thesis, *timed residuals* $TRes$ are proposed, which are used to isolate faults based on the time fault symptoms *early, late, and deadlock behavior*. The timed FDI approach is initially presented with a timed monolithic FDI model. After that, the approach is extended to distributed FDI, based on a timed distributed diagnosis model, and is finally applied to the BMS.

5.2 Evaluation

The timed fault detection and isolation concept relies on a state estimation of the applied timed model and on the determination of time constraints. These computations are performed by the *evaluator* before the actual FDI task is carried out. Therefore, the evaluator contains a timed model that represents the logical and time fault-free behavior of the DES. The model used in this FDI framework is a *TAAO* introduced in Chapter 3. The state estimation is performed according to the following equations:

$$\widetilde{X}_t = \{x \in X \mid \lambda(x) = u(t)\} \tag{5.1}$$

and

$$\begin{aligned}\widetilde{X}_t = \{x \in X \mid &(\exists x_{pre} \in \widetilde{X}_{t-1}) \wedge ((x_{pre}, [\tau_{lo}, \tau_{up}], c, x) \in TT) \wedge \\ &(\lambda(x) = u(t)) \wedge (c(t) \in [\tau_{lo}, \tau_{up}])\}\end{aligned} \tag{5.2}$$

with $u(t)$ as the system output of the DES, $\lambda(x)$ is the model output, $c(t)$ is the value of the clock c, and $\widetilde{X}_{t-1}$ is the set of former estimated states at time $t-1$.[14] The time variable t is used in the following to indicate that the state is continuously estimated with the progress of time. The estimation by Equation 5.1 is performed when no former estimated states are available $\widetilde{X}_{t-1} = \emptyset$. Based on the observed system output $u(t)$, each state $x \in X$ with the same modeled output $\lambda(x)$ is a valid current state. The set $\widetilde{X}_{t-1}$ is empty during start-up of the diagnosis and after detecting a fault at time $t-1$. If $\widetilde{X}_{t-1} \neq \emptyset$, state estimation is performed according to Equation 5.2. Besides the new observation $u(t)$, $\widetilde{X}_t$ depends on the former estimated states $\widetilde{X}_{t-1}$ and on $c(t)$ representing the time that has elapsed since the observation of the last new system output. If for any former estimated state $x_{pre} \in \widetilde{X}_{t-1}$ a following state $x \in X$ exists such that the output of x corresponds the new observation $u(t)$ and $c(t)$ is within the acceptable time interval $c(t) \in [\tau_{lo}, \tau_{up}]$, then the following state x is added to the new current state estimation set $\widetilde{X}_t$. In general, these conditions may hold for more than one following state. A state estimation is *unambiguous*, if $|\widetilde{X}_t| = 1$ holds.

Algorithm 9 Timed state estimation algorithm

Require: Timed automaton *TAAO*, current system output $u(t)$, former system output $u(t-1)$ and former state estimation $\widetilde{X}_{t-1}$

1: **if** $|\widetilde{X}_{t-1}| = 0$ **then**
2:     Initialize $\widetilde{X}_t$ according to Equation 5.1
3: **else if** $(|\widetilde{X}_{t-1}| > 0) \wedge (u(t) \neq u(t-1))$ **then**
4:     Determine next $\widetilde{X}_t$ according to Equation 5.2
5: **else**
6:     $\widetilde{X}_t = \widetilde{X}_{t-1}$
7: **end if**
8: **return** $\widetilde{X}_t$

The timed state estimation is implemented by Algorithm 9. The algorithm requires the *TAAO*, the current system output $u(t)$, the former system output $u(t-1)$, and the

[14]System output u and model output $\lambda(x)$ are represented as I/O-vectors in the following. Each element of an I/O-vector denotes the value of a particular I/O, respectively, see Section 3.4 in Chapter 3.

former state estimation $\widetilde{X}_{t-1}$. It is executed at each time step t to ensure a continuous fault diagnosis monitoring of the DES. If the set of former estimated states $\widetilde{X}_{t-1}$ is empty, $\widetilde{X}_t$ is determined in Line 2 by Equation 5.1. If $|\widetilde{X}_{t-1}| > 0$ and a new I/O-vector $u(t) \neq u(t-1)$ is observed at time t, then the state estimation is performed based on Equation 5.2. If $|\widetilde{X}_{t-1}| > 0$ and no new output is observed, then the set of estimated states remains unchanged.

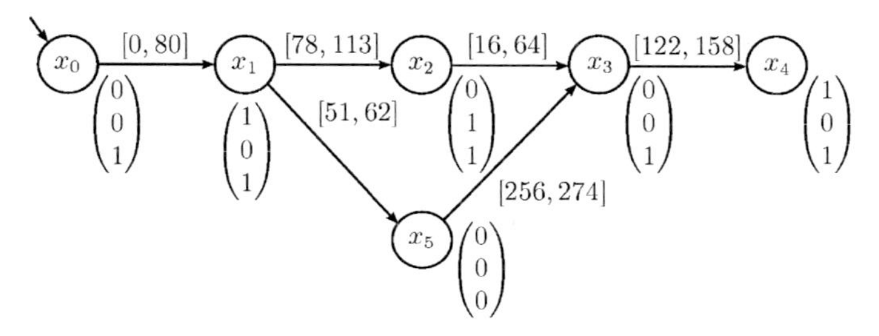

Figure 5.2: Example *TAAO*

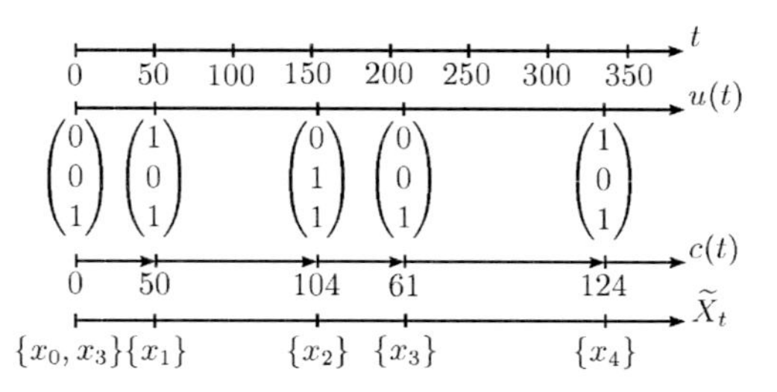

Figure 5.3: State estimation for an example sequence

Example 5.1 (Timed state estimation). The operation of the timed state estimation algorithm is illustrated by an example. In Figure 5.2, an identified, monolithic *TAAO* is shown that is used by the evaluator. Figure 5.3 shows an example sequence of I/O-vectors that are observed during the fault-free operation of the modeled DES. In the initial step $t = 0$, all states $x \in X$ are added to $\widetilde{X}_t$ for which $\lambda(x) = u(t)$. This holds in the example for x_0 and x_3 since $u(t) = \lambda(x_0) = \lambda(x_3)$. The state estimation for the following time is then based on $\widetilde{X}_{t-1} = \{x_0, x_3\}$. From $t = 1$ to $t = 49$, the set of estimated states remains unchanged $\widetilde{X}_t = \widetilde{X}_{t-1}$ since no new I/O-vector $u(t)$ is observed. At time $t = 50$, $u(t) \neq u(t-1)$ and $c(t) = 50$ holds. Based on $\widetilde{X}_{t-1}$, the following possible transitions have to be checked: $(x_0, [0, 80], c, x_1)$ with $u(t) = \lambda(x_1)$ and $c(t) \in [0, 80]$, and $(x_3, [122, 158], c, x_4)$ with $u(t) = \lambda(x_4)$ and $c(t) \notin [122, 158]$. Since the logical and time condition are only fulfilled for the transition from x_0 to the following state x_1, the current state estimation results in $\widetilde{X}_t = \{x_1\}$. This state estimation is unambiguous since $|\widetilde{X}_t| = 1$. After determination of a new set of estimated states based on a new observed system output, the clock $c(t)$ is reset. This is indicated by restarting arrows in Figure 5.3. For the following observations, the state estimations are performed accordingly.

The state estimation algorithm is an extension of the algorithm proposed in [Roth, 2010], which relies exclusively on the observed and modeled logical behavior. In that work, a theorem is proposed that if an automaton is logically precise, according to the logical precision discussion in Chapter 3, it is sufficient to observe a k-long sequence in order to determine an unambiguous state estimation $|\widetilde{X}_t| = 1$. Since an identified *TAAO* is logically precise with respect to k, this sufficient condition also holds for the timed state estimation algorithm. However, the observed and modeled time information can help to reduce the number of observations until an unambiguous state estimation is obtained. In Example 5.1, two observed output vectors are necessary in order to reach $|\widetilde{X}_t| = 1$ using the timed state estimation algorithm. If time information would not be used, then the conditions for state estimation after Equation 5.2 are reduced to the comparison of observed DES outputs $u(t)$ and model outputs $\lambda(x)$ for possible following states. Based on this, the untimed state estimation in the initial step $t = 0$ would lead to the same results as for the timed estimation. At time $t = 50$ the untimed state estimation would result in $\widetilde{X}'_t = \{x_1, x_4\}$ since $u(t) = \lambda(x_1) = \lambda(x_4)$. The distinction between x_1 and x_4 is only possible by additionally considering the time behavior. Based on untimed state estimation, one more observation would be necessary in the example until an unambiguous state can be determined.

Besides state estimation, the evaluator has to determine the *maximum sojourn time* that is necessary for deadlock detection. This is done according to the following equation:

$$\hat{\tau}_{up}(\widetilde{X}_t) = \max\{\tau_{up} \,|(\exists \widetilde{x} \in \widetilde{X}_t) \colon ((\widetilde{x}, [\tau_{lo}, \tau_{up}], c, x') \in TT)\}. \tag{5.3}$$

with $\widetilde{X}_t$ as the set of currently estimated states. The *maximum sojourn time* $\hat{\tau}_{up}(\widetilde{X}_t)$ is the largest τ_{up} of all outgoing transitions of $\widetilde{x} \in \widetilde{X}_t$. If either $\widetilde{X}_t = \emptyset$ or $\exists(\widetilde{x}, [\tau_{lo}, \tau_{up}], c, x') \in TT$ with $\tau_{up} = \infty$, then $\hat{\tau}_{up}(\widetilde{X}_t) = \infty$.

5.3 Timed Fault Detection

Timed fault detection relies on the information provided by the evaluator, which is the set of estimates states $\widetilde{X}_t$, the clock value $c(t)$ at time t, when the new observation $u(t)$ is made, and the maximum sojourn time $\hat{\tau}_{up}(\widetilde{X}_t)$. A fault is detected if an observation is not consistent with the modeled behavior. The fault detection policy is given as:

$$FD(\widetilde{X}_t, c(t), \hat{\tau}_{up}(\widetilde{X}_t)) = \begin{cases} ok & \text{if } (|\widetilde{X}_t| = 1) \wedge (c(t) \leq \hat{\tau}_{up}(\widetilde{X}_t)) \\ f_{init} & \text{if } (|\widetilde{X}_t| > 1) \wedge (c(t) \leq \hat{\tau}_{up}(\widetilde{X}_t)) \\ f_d & \text{if } (|\widetilde{X}_t| \neq 0) \wedge (c(t) > \hat{\tau}_{up}(\widetilde{X}_t)) \\ f & \text{if } (|\widetilde{X}_t| = 0) \end{cases} \tag{5.4}$$

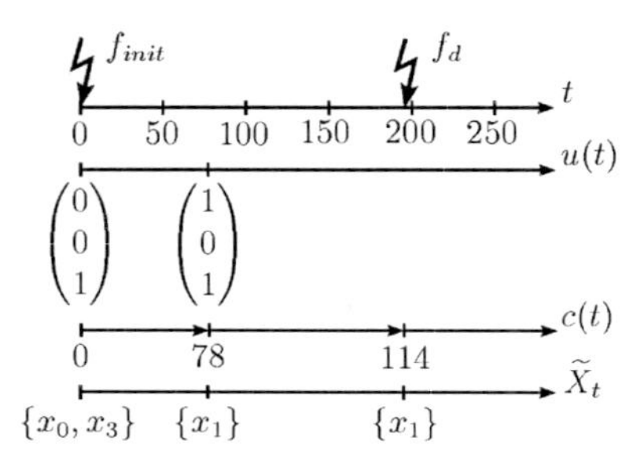

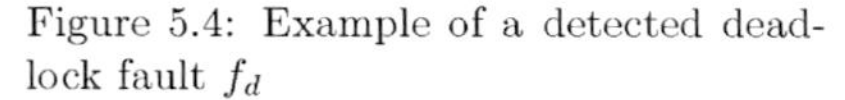

Figure 5.4: Example of a detected deadlock fault f_d

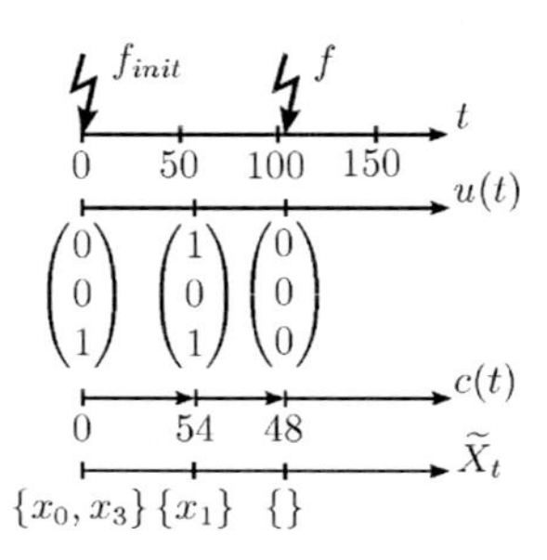

Figure 5.5: Example of a detected fault f due to time inconsistency

with *ok* indicating the fault-free operation of the system, f_{init} is the initialization fault, f_d is the deadlock fault, and f is the logical and time fault. Timed fault detection is continuously performed after timed state estimation and the determination of the maximum sojourn time. In the case that the current model state is unambiguously estimated $|\widetilde{X}_t| = 1$ and the clock value is below the maximum sojourn time $c(t) \leq \hat{\tau}_{up}(\widetilde{X}_t)$, then it is concluded that the system is operating fault-free. This is indicated by the *ok* label. An initialization fault f_{init} is detected if the state estimation is ambiguous after initialization of the model while $c(t)$ does not exceed the maximum sojourn time $\tau_{up}(\widetilde{X}_t)$ for the estimated states $\widetilde{X}_t$. The time condition allows for deadlock detection although an unambiguous state is not determined yet. In general, a deadlock fault f_d is detected if $\widetilde{X}_t$ contains at least one state and the clock $c(t)$ exceeds the maximum sojourn time $\tau_{up}(\widetilde{X}_t)$. This means that for $c(t)$, no outgoing transition of the estimated states $\widetilde{x} \in \widetilde{X}_t$ exists such that the time condition, given by $guard(\widetilde{x}, x')$, will be satisfied. Since no transition can be executed within the model any longer, the system is considered as deadlocked. If no state can be estimated on the occurrence of a new system output, a fault f is detected. This fault represents the cases when no following state x' can be found such that $u(t) = \lambda(x')$ and when a following state can be found but the time condition is violated $c(t) \notin guard(\widetilde{x}, x')$. For fault detection it is not necessary to distinguish between the logical and time cause of f. This will be done during fault isolation. Nevertheless, the distinction of f_d and f is important since they require different fault isolation strategies. The detection of a deadlock fault f_d is *model-triggered*. It relies on the continuous update of the time information $c(t)$ that is provided by the model. Only modeled information will be used for fault isolation. In contrast to that, the detection of a fault f is *DES-triggered* upon the generation of a new system output $u(t)$. In this case, fault isolation is performed by means of the modeled and observed behavior. The faults that are relevant for fault isolation are f and f_d. Since an initialization fault f_{init} does not refer to a faulty system component, fault isolation will not be executed.

Example 5.2. In Figure 5.4 to Figure 5.7, four example output sequences are presented in which faults are detected based on the *TAAO*, given in Figure 5.2. Given the output

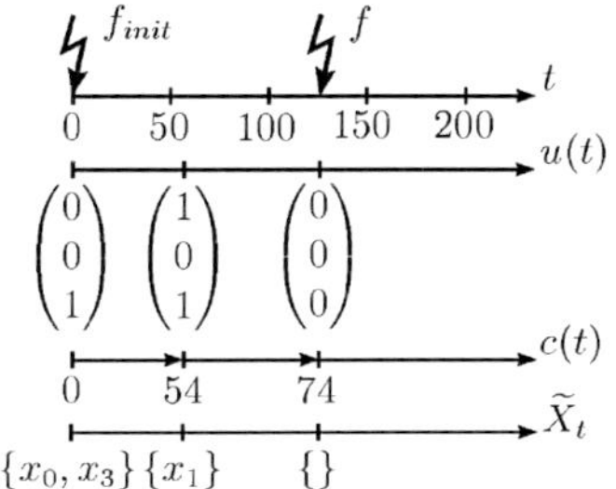

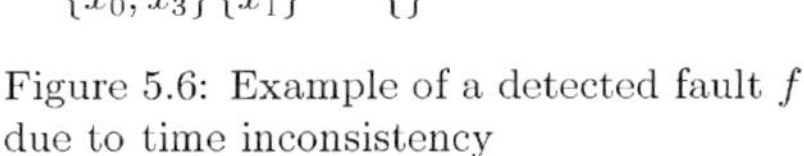

Figure 5.6: Example of a detected fault f due to time inconsistency

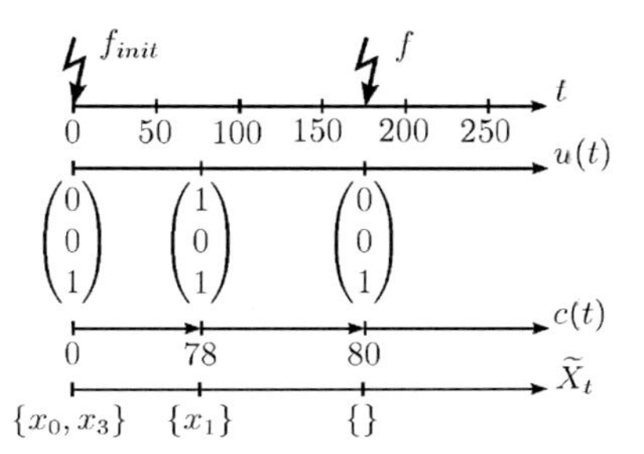

Figure 5.7: Example of a detected fault f due to logical inconsistency

sequence in Figure 5.4, a deadlock fault f_d is detected. At time $t = 78$, state x_1 is unambiguously estimated. Based on $\widetilde{X}_t = \{x_1\}$, two outgoing transitions $(x_1, [78, 113], c, x_2)$ and $(x_1, [51, 62], c, x_5)$ exist with maximum sojourn time $\hat{\tau}_{up}(\widetilde{X}_t) = 113$. Since at time $t = 192$, $|\widetilde{X}_t| \neq 0$ and $c(t) = 114 > \hat{\tau}_{up}(\widetilde{X}_t)$ holds, the occurrence of a deadlock fault f_d is concluded. The second and third sequence presented in Figures 5.5 and 5.6 lead to the detection of the fault f since a new system output $u(t)$ is observed that satisfies the logical condition but violates the modeled time constraints, respectively. The state estimation for both observed sequences yield $\widetilde{X}_t = \{x_1\}$ at time $t = 54$. The two possible outgoing transitions have the target states x_2 and x_5. Since $u(t) = \lambda(x_5)$ holds for both observed output sequences, based on the next observed system output, the logical condition of the next state estimation is fulfilled. However, the time condition is not since $c(t) = 48 \notin [51, 62]$ in Figure 5.5 and $c(t) = 74 \notin [51, 62]$ in Figure 5.6. Hence, the result of the next state estimation is $\widetilde{X}_t = \emptyset$ and fault f is detected. The fourth example sequence, leading to a detection of f, based on a logical misbehavior, is depicted in Figure 5.7. After estimating $\widetilde{X}_t = \{x_1\}$, a new system output $u(t)$ is observed at absolute time $t = 158$ that is logically inconsistent with the outputs of the possible following states $\lambda(x_2) \neq u(t)$ and $\lambda(x_5) \neq u(t)$. As a result, the state estimation is $\widetilde{X}_t = \emptyset$ and the fault detection determines $FD = f$.

It has been discussed in Chapter 2 that identified models used for FDI purposes need to be precise and complete in order to avoid missed and false detections. An imprecise model contains exceeding timed behavior $L_{EXC,t}$ that may correspond to behavior produced by the DES upon the occurrence of a fault. In this case, this specific fault will be missed by the fault detection since the model is able to reproduce the observed fault induced behavior. In particular, an unambiguous timed state estimation $\widetilde{X}_t$ will be determined for all observed system outputs $u(t)$ after the occurrence of the fault and all clock values $c(t)$ will satisfy the related maximum sojourn time $\tau_{up}(\widetilde{X}_t)$. In order to avoid logical and time related false detections, the applied FDI model is supposed to be as precise as possible. False detections refer to an incompleteness of the applied FDI model. If a model is incomplete, non-reproducible timed behavior $L_{NR,t}$ exists that corresponds to the fault-free system behavior. If non-reproducible behavior is observed, it cannot be reproduced the model and a fault is falsely detected. In particular, given an unambiguous timed state estimation $\widetilde{X}_t$, a new observed, fault-free system output $u(t)$ leads to $|\widetilde{X}_t| = 0$ or $c(t) > \hat{\tau}_{up}(\widetilde{X}_t)$ such that either a deadlock fault f_d or a logical and time fault f is declared. False detections can be avoided by using complete models for FDI. It can be seen that requirements placed for the acceptable number of missed and false detections need to be explicitly considered during the modeling procedure by means of the model precision and completeness. Models identified by use of the algorithms presented in Chapter 3 give guarantees for these properties, making them appropriate for fault detection purposes.

5.4 Timed Fault Isolation

5.4.1 Preliminaries

After detecting a fault, the failed system components need to be isolated in order to arrange for the required repairs. In this work, the aim of fault isolation is to determine a comparable small number of sensors and actuators that are related to the detected fault. This allows a maintenance crew to focus on a rather limited part of the DES, instead of a comprehensive and costly check of all system components. The presented isolation concept follows the principle proposed in [Reiter, 1987] in which a proper fault isolation is determined by assuming a minimum number of system components to be failed such that the observed faulty behavior can be explained. The principle is based on the fact that a simultaneous failure of fewer components is more likely than the simultaneous failure of many of them. From all possible explanations, which would explain the detected fault, the one which applies most assumes the smallest number of components to be failed. This holds for industrial systems in particular because reliable hardware is typically used that fails very rarely with respect to its overall mission time. In closed-loop DES, timed fault isolation is the determination of time fault symptoms based on observed and modeled timed system outputs. The time fault symptoms refer to time behavior of I/Os that is inconsistent with the modeled one. I/Os generating time fault symptoms are called fault candidates. Since each controller I/O is connected to a sensor or actuator in the DES, the fault candidates allow to immediately localize the fault within the plant.

Time fault symptoms of a closed-loop DES are determined based on the time dependent behavior of each single I/O. Given an I/O with binary value range and two discrete times j and k, the I/O can change its value from 0 to 1, denoted as *rising edge*, from 1 to 0, denoted as *falling edge*, or its value remains the same. These edges are determined by the following function.

Definition 35 (Edge function).

$$edge(io(j), io(k)) = \begin{cases} io\uparrow & \text{if } io(j) = 0 \text{ and } io(k) = 1 \\ io\downarrow & \text{if } io(j) = 1 \text{ and } io(k) = 0 \end{cases} \tag{5.5}$$

with a rising edge of io indicated by $io\uparrow$ and a falling edge indicated by $io\downarrow$.

Given a system output u as I/O-vector at discrete time steps j and k, the edges for all I/Os of u are derived according to the following function.

Definition 36 (I/O-vector edges).

$$vEdge(u(j), u(k)) = \bigcup_{i=1}^{m} \{edge\,(u(j)[i], u(k)[i])\} \tag{5.6}$$

in which $u(j)[i]$ and $u(k)[i]$ address the i-th I/O of $u(j)$ and $u(k)$, respectively.

More than one I/O of an I/O-vector u can change its value from j to k, in general. In order to determine the I/O-vector edges of modeled outputs, the system output u of Definition 36 can be replaced by the model output $\lambda(x)$.

Example 5.3 (I/O-vector edges). Two I/O-vectors $u(1)$ and $u(2)$ with $m = 3$ are given as

$$u(1) = \begin{pmatrix} 1 \\ 0 \\ 1 \end{pmatrix} \text{ and } u(2) = \begin{pmatrix} 1 \\ 1 \\ 0 \end{pmatrix}.$$

The I/O-vector edges are determined as $vEdge(u(1), u(2)) = \{io_2\uparrow, io_3\downarrow\}$.

5.4.2 Deadlock Behavior

For the isolation of a deadlock fault f_d, two timed residuals are proposed. Given the last unambiguously estimated state $\widetilde{x}$ before f_d is detected, the first *deadlock residual* is determined by the following equation.

$$TRes_d^{\cap}(\widetilde{x}) = \bigcap_{(\widetilde{x}, guard, c, x') \in TT} vEdge\,(\lambda(\widetilde{x}), \lambda(x')) \tag{5.7}$$

The residual contains the rising and falling edges of I/Os that were expected to occur within the admissible maximum sojourn time. These edges are supposed to occur for any following state, independent of the transition taken. The I/O-vector edges $vEdge\,(\lambda(\widetilde{x}), \lambda(x'))$ are determined based on the modeled output $\lambda(\widetilde{x})$ of last estimated state before the fault is detected and on the outputs of its possible following states $\lambda(x')$ that are connected with a transition $(x, guard, c, x') \in TT$. The intersection operation determines those edges that are supposed to occur upon the next transition, no matter which one is taken. The next deadlock residual is a less strict formulation of the first one. It determines edges that can occur for at least one following state.

$$TRes_d^{\cup}(\widetilde{x}) = \bigcup_{(\widetilde{x}, guard, c, x') \in TT} vEdge\,(\lambda(\widetilde{x}), \lambda(x')) \tag{5.8}$$

The I/O-vector edges $vEdge\,(\lambda(\widetilde{x}), \lambda(x'))$ are determined as for the first deadlock residual. Instead of using the intersection operation, the union operation is applied here. In that way, all rising and falling edges that can be generated for all following states of $\widetilde{x}$ are added. Obviously, $TRes_d^{\cup}(\widetilde{x}) \supseteq TRes_d^{\cap}(\widetilde{x})$ holds. A maintenance crew is advised to check the fault candidates of $TRes_d^{\cap}(\widetilde{x})$ first since their behavior is more likely connected to the detected fault. If a fault cannot be isolated by means of them, the remaining candidates given by the less strict deadlock residual $TRes_d^{\cup}(\widetilde{x})$ will be checked.

Example 5.4 (Deadlock behavior). Given the *TAAO* in Figure 5.2 and the observed sequence in Figure 5.4, in which a deadlock fault f_d is detected. The state $\widetilde{x} = x_1$ is unambiguously estimated before the fault is detected. Upon detection of f_d, the deadlock behavior residuals are determined as

$$TRes_d^{\cap}(x_1) = \{io_1\downarrow, io_2\uparrow\} \cap \{io_1\downarrow, io_3\downarrow\} = \{io_1\downarrow\}$$

and

$$TRes_d^{\cup}(x_1) = \{io_1\downarrow, io_2\uparrow\} \cup \{io_1\downarrow, io_3\downarrow\} = \{io_1\downarrow, io_2\uparrow, io_3\downarrow\}.$$

The first deadlock residual contains the single edge $io_1\downarrow$. This means that $io_1\downarrow$ was expected to occur in any case. If the fault candidate io_1 is not responsible for the fault, the remaining fault candidates provided by $TRes_d^{\cup}(x_1)$ are io_2 and io_3.

5.4.3 Early and Late Behavior

A fault f is detected if a new system output $u(j)$ is observed that is logically or temporally inconsistent with the modeled behavior. In order to isolate f in case of a logical inconsistency, the logical residuals for unexpected and missed behavior from [Roth, 2010] can be applied. In the following, two timed residuals are presented that focus on the time fault symptoms *early and late behavior*. Time fault symptoms are generated if the observed behavior is consistent with the modeled logical behavior but inconsistent with the modeled time behavior. The *early behavior residual* is given as:

$$\begin{aligned} TRes_{eb}(\widetilde{x}, u(t), c(t)) = \{vEdge(\lambda(\widetilde{x}), u(t)) \mid (\exists(\widetilde{x}, [\tau_{lo}, \tau_{up}], c, x') \in TT) \wedge \\ (u(t) = \lambda(x')) \wedge (c(t) < \tau_{lo})\} \end{aligned} \tag{5.9}$$

with $\widetilde{x} \in \widetilde{X}$ as the last unambiguously estimated state before detection of f, $u(t)$ is the system output that lead to fault detection, and $c(t)$ is the corresponding clock value. The early behavior residual contains I/O-vector edges $vEdge(\lambda(\widetilde{x}), u(t))$ that are generated upon the early reception of a new logically consistent system output $u(t)$. The notation 'early' means that the relative time when $u(t)$ is observed is smaller than expected with respect to the modeled time guards. Therefore, $vEdge(\lambda(\widetilde{x}), u(t))$ determines the set of rising and falling edges between the estimated model state before fault detection $\widetilde{x}$ and the observed system output $u(t)$ that led to fault detection. The edges are part of the residual if a following state x' of $\widetilde{x}$ exists with $u(t) = \lambda(x')$ and the relative time passed between activation of $\widetilde{x}$ and fault detection is below the lower time bound $c(t) < \tau_{lo}$ of the related transition. The *late behavior residual* is defined likewise by the following equation.

$$\begin{aligned} TRes_{lb}(\widetilde{x}, u(t), c(t)) = \{vEdge(\lambda(\widetilde{x}), u(t)) \mid (\exists(\widetilde{x}, [\tau_{lo}, \tau_{up}], c, x') \in TT) \wedge \\ (u(t) = \lambda(x')) \wedge (c(t) > \tau_{up})\} \end{aligned} \tag{5.10}$$

The late behavior residual is the counterpart of the early behavior residual. It contains I/O-vector edges $vEdge(\lambda(\widetilde{x}), \lambda(x'))$ that are consistent with the logical, modeled behavior but observed late with respect to the modeled time bounds. If a system output $u(t)$ is observed such that $\nexists(\widetilde{x}, [\tau_{lo}, \tau_{up}], c, x') \in TT$ with $u(t) = \lambda(x')$, then $TRes_{eb}(\widetilde{x}, u(t), c(t)) = \emptyset$ and $TRes_{lb}(\widetilde{x}, u(t), c(t)) = \emptyset$. If a fault f is detected and the observation is logically inconsistent with the model, then the timed residuals will return no fault candidates.

Example 5.5 (Early and late behavior). Given the $TAAO$ in Figure 5.2 and the observed output sequence in Figure 5.5, related to the detected fault f at time $t = 102$. Before the system output

$$u(t) = \begin{pmatrix} 0 \\ 0 \\ 0 \end{pmatrix}$$

is observed and $c(t) = 48$, $\widetilde{x} = x_1$ represents the last valid unambiguously estimated state. Since $u(t) = \lambda(x_5)$, the observed behavior is logically consistent with the model and the early and late behavior residuals are determined as

$$\begin{aligned} TRes_{eb}(x_1, u(t), 48) &= \{io_1\downarrow, io_3\downarrow\}, \\ TRes_{lb}(x_1, u(t), 48) &= \emptyset. \end{aligned}$$

The second observed sequence is given in Figure 5.6, in which again a fault f is detected. As before, the state $\widetilde{x} = x_1$ is the last valid state and the same logically consistent controller I/O-vector $u(t)$ is observed. The fault is now detected at $c(t) = 74$ and the early and late behavior residuals result in

$$\begin{aligned} TRes_{eb}(x_1, u(t), 74) &= \emptyset, \\ TRes_{lb}(x_1, u(t), 74) &= \{io_1\downarrow, io_3\downarrow\}. \end{aligned}$$

The number of fault candidates determined by the residuals depends in general on the number of possible following states of an estimated state and on the number of I/Os that change their value simultaneously upon a transition. The total number of I/Os of a closed-loop DES has no influence on that. This is important, since it allows that even for large systems a rather small number of fault candidates can be obtained that allows for a fast localization of the failed hardware components. It should be further remarked that the reason for a system fault is not necessarily the failure of a sensor or actuator. Any hardware component in the DES may fail such that the system is unavailable. However, the failure of a hardware component that contributes to the production process causes associated sensors and actuators to behave in an abnormal way. This leads to the generation of fault symptoms which can be recognized and interpreted by the FDI system. Although these components cannot be immediately isolated by the proposed FDI approach, by isolating the abnormal behaving sensors and actuators a part of the DES can be determined where the actual failed hardware is most likely located.

The proposed two-dimensional fault isolation concept for DES with logical and timed residuals is summarized in Figure 5.8. The logical dimension is given by the residuals for the logical fault symptoms unexpected and missed behavior. The timed residuals, on

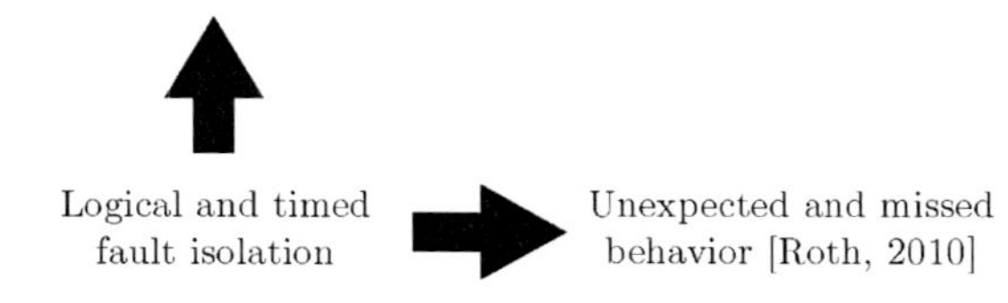

Figure 5.8: Two-dimensional fault isolation with DES residuals

the other hand, focus on the time fault symptoms deadlocks, early, and late behavior. They represent the second, timed dimension of fault isolation. The two-dimensional fault isolation concept reflects the idea of the distinction between logical and time behavior of the DES, as it has been pursued during modeling and identification in Chapter 3 of this thesis. To derive the timed residuals, logical consistency of an observation is a necessary condition. If an observation is logically inconsistent with the modeled behavior, the detected fault f will refer to a logical misbehavior of the system and the timed residuals will not be determined. In that way, fault isolation draws the distinction between logical and time related faults.

5.5 Extension to Timed Distributed Models

5.5.1 Overview

The presented procedures for fault detection and isolation require a monolithic model of the considered DES. In order to use timed distributed models for FDI, a *distributed fault diagnosis* is presented in the following. Recall that a timed distributed model is given as $TDM = \{TAAO_1, TAAO_2, \ldots, TAAO_N\}$ with $TAAO_i$ denoting the partial time automaton that models the i-th subsystem of the DES with respect to the partial system output $u_i(t)$. The applied timed distributed modeling framework and the procedures for timed identification are described in detail in Chapter 3. In Figure 5.9, an overview of the distributed FDI concept is given. The output of the DES is online observed, like in the monolithic case. The differences are given with the evaluation and the FDI procedures. The distributed evaluation performs state estimation and maximum sojourn time determination for all partial automata of the TDM. The model output represents all N distributed evaluation results here. Based on the DES and determined model output, *local FDI* calculations are performed for all subsystem. The local FDI includes all calculations for fault detection and isolation with respect to the partial observed and modeled behavior of the considered subsystem. The local results are combined by *global FDI* to the global fault detection FD_{TDM} and the global timed residuals $TRes_{TDM}$.

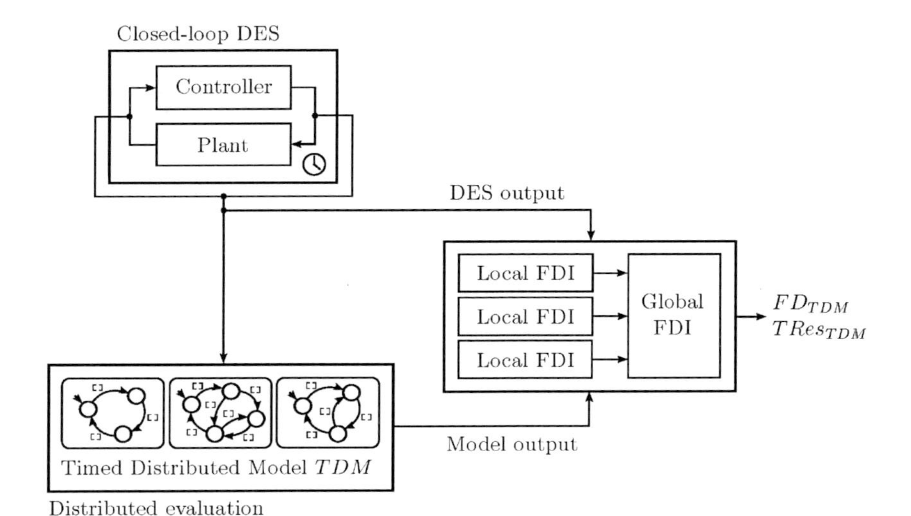

Figure 5.9: Overview of model-based FDI using a timed distributed model

5.5.2 Distributed Evaluation

A TDM is evaluated by performing an individual evaluation for each partial automaton, respectively. The evaluation procedures have to be executed N times such that for each $TAAO_i \in TDM$ a state estimation $\widetilde{X}_{i,t}$ and a belonging maximum sojourn time $\hat{\tau}_{up}(\widetilde{X}_{i,t})$ is determined. State estimation for a partial automaton $TAAO_i$ is performed based on the partial system output $u_i(t)$. The following algorithm describes the distributed timed state estimation procedure.

Algorithm 10 Timed distributed state estimation algorithm

Require: Timed distributed model $TDM = \{TAAO_1, TAAO_2, \ldots, TAAO_N\}$, current system output $u(t)$, former system output $u(t-1)$, former state estimations $\widetilde{X}_{TDM,t-1} = (\widetilde{X}_{1,t-1}, \widetilde{X}_{2,t-1}, \ldots, \widetilde{X}_{N,t-1})$

1: **for all** $TAAO_i \in TDM$ **do**
2: Determine the partial system outputs $u_i(t)$ and $u_i(t-1)$ using Definition 22
3: Determine the partial state estimations $\widetilde{X}_{i,t}$ according to Algorithm 9
4: **end for**
5: **return** $\widetilde{X}_{TDM,t} = (\widetilde{X}_{1,t}, \widetilde{X}_{2,t}, \ldots, \widetilde{X}_{N,t})$

The timed distributed state estimation algorithm requires the partial automata given by the timed distributed model TDM, the currently and formerly observed output vectors $u(t)$ and $u(t-1)$ of the DES, and the former state estimation $\widetilde{X}_{i,t-1}$ for each $TAAO_i \in$

TDM. At time t, the state estimation is performed individually for each $TAAO_i \in TDM$ in the for-loop from Line 1 to Line 4 of the algorithm. First, the current and former partial system outputs $u_i(t)$ and $u_i(t-1)$ are determined according to Definition 22 based on $u(t)$ and $u(t-1)$. The partial system outputs $u_i(t)$ and $u_i(t-1)$ respect those I/Os of $u(t)$ and $u(t-1)$ that are related to the i-th subsystem. Then, the partial state estimation for $TAAO_i$, according to Algorithm 9, is performed. Each local estimation relies on the partial model $TAAO_i$, the local automaton clock $c_i(t)$, the determined partial system outputs $u_i(t)$ and $u_i(t-1)$, and the former state estimation $\widetilde{X}_{i,t-1}$. Finally, the collection of estimated states for the TDM given by $\widetilde{X}_{TDM,t} = (\widetilde{X}_{1,t}, \widetilde{X}_{2,t}, \ldots, \widetilde{X}_{N,t})$ is returned by the algorithm. The determination of the maximum sojourn time $\hat{\tau}_{up}(\widetilde{X}_{i,t})$ for the i-th automaton based on the current local state estimation $\widetilde{X}_{i,t}$ is done according to Equation 5.3. An example for timed distributed state estimation will be given after the presentation of the global fault detection and isolation approach.

5.5.3 Distributed Fault Detection and Isolation

The idea of fault diagnosis using a timed distributed model is to perform local fault diagnosis for each subsystem of the DES and to combine the local results to the *global fault detection and isolation result.* The global FDI result is given by the global fault detection FD_{TDM} and the global timed residuals $TRes_{TDM}$. The procedure to determine these results is rather a decision than a calculation since no additional fault detection or isolation computations are made on the global level. The local FDI calculations are performed according to the monolithic FDI procedure, in the sense that each partial automaton is treated as a 'monolithic model' of the considered subsystem. For global fault detection, the following decision policy is given:

$$FD_{TDM} = \begin{cases} ok & \text{if } (\forall TAAO_i \in TDM)\colon FD(\widetilde{X}_{i,t}, c_i(t), \hat{\tau}_{up}(\widetilde{X}_{i,t})) = ok \\ f_{init} & \text{if } (\exists TAAO_i \in TDM)\colon FD(\widetilde{X}_{i,t}, c_i(t), \hat{\tau}_{up}(\widetilde{X}_{i,t})) = f_{init} \\ f_d & \text{if } (\exists TAAO_i \in TDM)\colon FD(\widetilde{X}_{i,t}, c_i(t), \hat{\tau}_{up}(\widetilde{X}_{i,t})) = f_d \\ f & \text{if } (\exists TAAO_i \in TDM)\colon FD(\widetilde{X}_{i,t}, c_i(t), \hat{\tau}_{up}(\widetilde{X}_{i,t})) = f \end{cases} \quad (5.11)$$

with $FD(\widetilde{X}_{i,t}, c_i(t), \hat{\tau}_{up}(\widetilde{X}_{i,t}))$ representing the local fault detection performed for $TAAO_i$, according to Equation 5.4. A DES is in the global *ok*-state, if all partial automata are in the *ok*-state. If at least one fault is reported by any local fault detection unit, then the global system also enters a fault state. This respects the fact that a fault may affect the behavior of a selection of subsystems only. In that case, the fault may be detected by any of the local units, while the remaining units report the system as operating fault-free. However, from the global point of view, the system has to be considered as faulty. Upon detection of a fault, the global fault detection decision FD_{TDM} may contain different fault states at the same time, for instance $FD_{TDM} = \{f_d, f\}$. This is reasonable since the occurrence of a fault may provoke different partial automata to produce different fault symptoms.

Deadlock Behavior

When a deadlock fault is detected by any of the local detection units $f_d \in FD_{TDM}$, the global deadlock residuals are determined. Initially, the partial automata have to be selected for which the deadlock fault is detected. This is done according to the equation

$$AUT_{f_d} = \{TAAO_i \mid FD(\widetilde{X}_{i,t}, c_i(t), \hat{\tau}_{up}(\widetilde{X}_{i,t})) = f_d\} \tag{5.12}$$

while $\forall TAAO_i \in AUT_{f_d}$, $|\widetilde{X}_{i,t}| = 1$ holds before fault detection. In general, $AUT_{f_d} \subseteq TDM$. The global deadlock residuals are then determined by the following two equations:

$$TRes^{\cap}_{TDM,d} = \bigcap_{TAAO_i \in AUT_{f_d}} TRes^{\cap}_{i,d}(\widetilde{x}_i) \tag{5.13}$$

and

$$TRes^{\cup}_{TDM,d} = \bigcup_{TAAO_i \in AUT_{f_d}} TRes^{\cup}_{i,d}(\widetilde{x}_i) \tag{5.14}$$

with $TRes^{\cap}_{i,d}$ and $TRes^{\cup}_{i,d}$ denoting the deadlock residuals according to Equations 5.7 and 5.8 specific to $TAAO_i$ and the locally estimated state $\widetilde{x}_i \in \widetilde{X}_{i,t}$. The residuals are determined for all $TAAO_i$ for which a fault f_d has been detected. The global deadlock residuals intersects the local results in case of $TRes^{\cap}_{TDM,d}$ and unifies them in case of $TRes^{\cup}_{TDM,d}$. The idea behind this is the same as for the local residuals. Edges of I/Os that are determined in all local residuals are more likely related to the fault, than edges which are isolated in only some of the considered partial automata.

Early and Late Behavior

Next, the global timed residuals for early and late behavior are introduced. They are determined if a fault $f \in FD_{TDM}$ is locally detected within any subsystem of the DES. As for the global deadlock residuals, the partial automata have to be determined in which the fault f is detected.

$$AUT_f = \{TAAO_i \mid FD(\widetilde{X}_{i,t}, c_i(t), \hat{\tau}_{up}(\widetilde{X}_{i,t})) = f\} \tag{5.15}$$

while $\forall TAAO_i \in AUT_f$, $|\widetilde{X}_{i,t}| = 1$ holds before fault detection. In general, $AUT_f \subseteq TDM$. The global early behavior residual is determined by the following equation:

$$TRes_{TDM,eb} = \bigcup_{TAAO_i \in AUT_f} TRes_{i,eb}(\widetilde{x}_i, u_i(t)) \tag{5.16}$$

with $TRes_{i,eb}$ denoting the early behavior residual according to Equation 5.9 specific to $TAAO_i$, to the locally estimated state $\widetilde{x}_i \in \widetilde{X}_{i,t}$, and to the partial system output $u_i(t)$. The global late behavior residual is determined by the following equation:

$$TRes_{TDM,lb} = \bigcup_{TAAO_i \in AUT_f} TRes_{i,lb}(\widetilde{x}_i, u_i(t)) \tag{5.17}$$

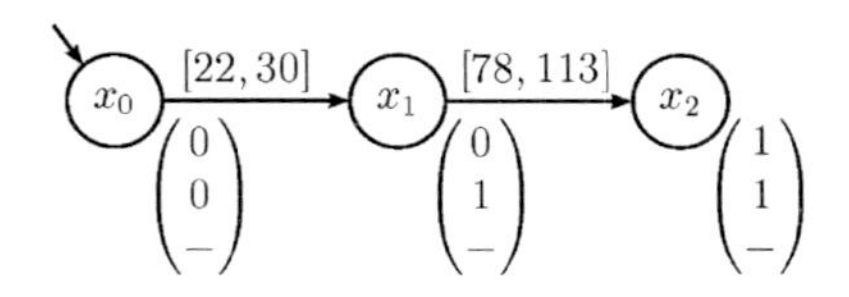

Figure 5.10: Example partial $TAAO_1$

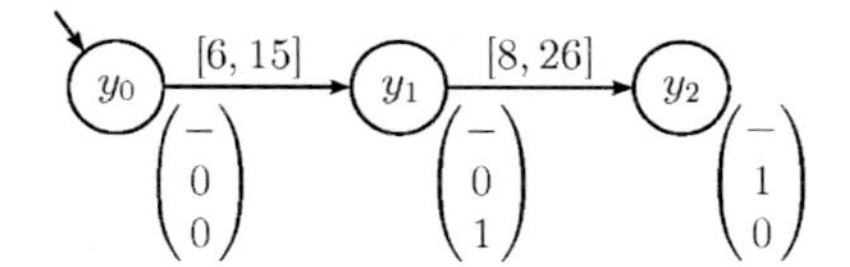

Figure 5.11: Example partial $TAAO_2$

Local FDI for $TAAO_1$	Local FDI for $TAAO_2$
$FD(\widetilde{X}_{1,t}, c_1(t), \hat{\tau}_{up}(\widetilde{X}_{1,t})) = f$	$FD(\widetilde{X}_{2,t}, c_2(t), \hat{\tau}_{up}(\widetilde{X}_{2,t})) = f$
$TRes_{eb}(x_0, u_1) = \{io_2\uparrow\}$	$TRes_{eb}(y_1, u_2) = \{io_2\uparrow, io_3\downarrow\}$
$TRes_{lb}(x_0, u_1) = \emptyset$	$TRes_{lb}(y_1, u_2) = \emptyset$

Global FDI
$FD_{TDM} = f$
$TRes_{TDM,eb} = \{io_2\uparrow, io_3\downarrow\}$
$TRes_{TDM,lb} = \emptyset$

Table 5.1: Local and global FDI results for σ_f

with $TRes_{i,lb}$ denoting the early behavior residual according to Equation 5.10 specific to $TAAO_i$, to the locally estimated state $\widetilde{x}_i \in \widetilde{X}_{i,t}$, and to the partial system output $u_i(t)$. The global timed residuals for early and late behavior contain all edges of the corresponding local determined residuals. It is not reasonable to draw a distinction between edges that are contained in all $TRes_{i,lb}(\widetilde{x}_i, u_i(t))$ or in less than all of them by using the intersection operation. I/Os producing edges such that an early or late fault is detected do all belong to the set of fault candidates.

Example 5.6 (Fault detection and isolation using a timed distributed model). This example illustrates the fault detection and isolation procedure using a timed distributed model. In Figure 5.10 and Figure 5.11, two partial automata $TAAO_1$ and $TAAO_2$ are depicted that constitute the timed distributed model $TDM = \{TAAO_1, TAAO_2\}$. The partial I/O-vectors, representing the partial system outputs of both partial automata, are given as

$$u_1 = \begin{pmatrix} io_1 \\ io_2 \\ - \end{pmatrix} \text{ and } u_2 = \begin{pmatrix} - \\ io_2 \\ io_3 \end{pmatrix}.$$

Assume that the corresponding is faulty and the following output sequence is observed:

$$\sigma_f = \left(\begin{pmatrix} 0 \\ 0 \\ 0 \end{pmatrix}, \begin{pmatrix} 0 \\ 0 \\ 1 \end{pmatrix}, \begin{pmatrix} 0 \\ 1 \\ 0 \end{pmatrix} \right).$$

In Figure 5.12 and Figure 5.13, the resulting partial system outputs $u_i(t)$, local clocks $c_i(t)$, and local state estimations $\widetilde{X}_{i,t}$ are shown. One can see that a fault f is locally

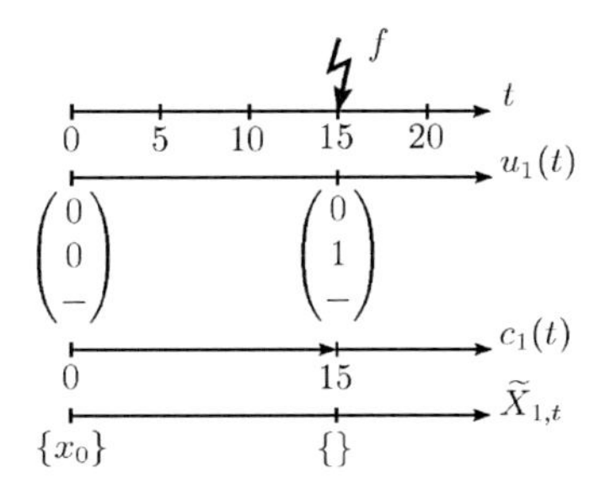

Figure 5.12: Local fault detection via $TAAO_1$

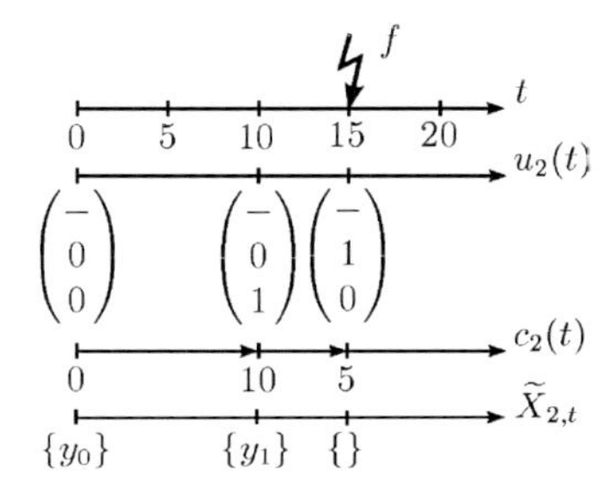

Figure 5.13: Local fault detection via $TAAO_2$

detected at $t = 15$, simultaneously by both local fault detection units. The observed partial system outputs are logically consistent with the partial automata but the time constraints $guard(x_0, x_1)$ and $guard(y_1, y_2)$ are violated, respectively. As a result, the early and late residuals are locally determined for both partial automata and partial system outputs. They show that io_2 is observed early with respect to $TAAO_1$ and io_2, io_3 are observed early with respect to $TAAO_2$. These local results are then combined to the global FDI result, that a fault f is detected based on an early occurrence of io_2 and io_3. The local and global FDI results of this example are summarized in Table 5.1.

5.6 Fault Detection and Isolation of the BMS

5.6.1 Online Fault Diagnosis Implementation

The presented FDI approach, using timed distributed models, was applied to the BMS. The evaluation system has been described in Chapter 1. In Figure 5.14, the developed online FDI implementation is shown. The distributed fault diagnosis algorithm and a timed distributed model of the controlled BMS were implemented on a PC. In order to online observe the BMS behavior during operation, the PC was linked with the PLC via the Modbus TCP connection. Controller I/O-vectors were periodically sent by the PLC to the PC. The procedures, implemented on the PLC for the transmission of controller I/O-vectors, did not affect the controlled operation of the BMS. This satisfies the requirement for a passive observation principle. The FDI reports, generated by the diagnosis system, are based on the global FDI results. A FDI report includes the information about the type of detected fault, the isolated I/Os, a description of the corresponding sensor or actuator component including its location in the BMS, and a time stamp when the fault is detected. The FDI reports were stored in a data-base that ran on the PC. Maintenance crews could access the diagnosis reports via a TCP/IP connected web-server using either operator workstations or mobile devices.

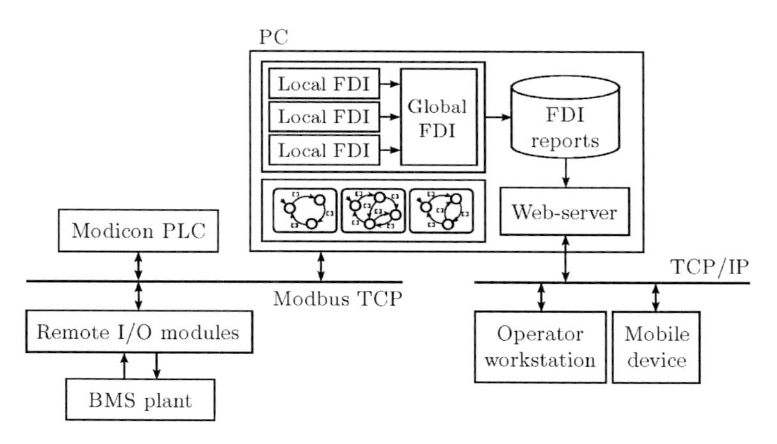

Figure 5.14: On-line fault diagnosis implementation of the BMS

5.6.2 Fault Scenarios

In order to demonstrate the fault detection and isolation capabilities of the proposed methods, the diagnosis of three different fault scenarios of the BMS will be discussed in the following. The considered fault scenarios are the *F1 stuck-on piston fault*, the *F2 blocked throttle valve fault*, and the *F3 worn-out test cylinder fault*. The BMS has been provoked to enter these fault states by manipulating the fault related hardware components. In Figure 5.15, these hardware components are shown. Their task within the manufacturing process and their associated fault-free and faulty behavior will be explained in the following. The BMS is a large concurrent system with $|IO| = 73$, for which no appropriate information about the behavior of plant and controller algorithm for

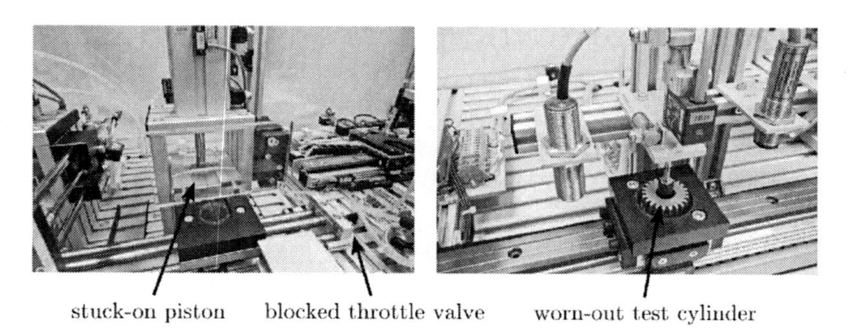

Figure 5.15: Fault scenarios of the BMS

FDI purposes was available. Consequently, the FDI model was automatically generated using the identification and partitioning approaches proposed in Chapters 3 and 4. The applied model was a TDM with $N = 10$ partial automata identified with respect to the logical identification parameter $k = 2$ and the time identification parameter $1 - \alpha = 0.99$, $\beta = 0.99$, and $ext_0 = 0.4$. For automatic modeling, $|\Sigma| = 50$ observations of the fault-free system behavior were used. All times presented in the following are given in milliseconds.

F1 Stuck-on Piston Fault

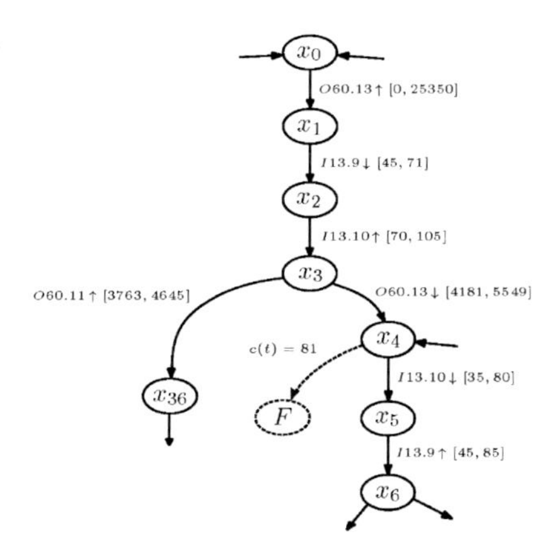

Figure 5.16: Detection of the stuck-on piston fault via $TAAO_3$

The first fault refers to the out-press module of the BMS located at station 3. During operation of the system, the out-press removed bearings from delivered workpieces. The fault-free behavior of the out-press was modeled by means of the partial automaton $TAAO_3$, depicted in Figure 5.16. For a proper visualization of the timed automaton, the model outputs $\lambda(x)$ are omitted in the figure. Instead, the corresponding I/O-vector edges $vEdge(\lambda(x), \lambda(x'))$ are depicted at the transitions in addition to the time guards. The operation of the press started with the delivery of a workpiece by the transportation unit. In this system state, represented by x_0 in $TAAO_3$, the piston of the cylinder was in its upper position. After a workpiece was inserted into the press, the piston was extended $O60.13\uparrow$. It left its upper position, indicated by $I13.9\downarrow$, and moved down until the lower position was reached $I13.10\uparrow$. The piston remained in this position for a few seconds to ensure that the bearing was pressed out. Then, the pressure from the cylinder was released $O60.13\downarrow$, the piston left the lower position $I13.10\downarrow$, and moved towards its upper position $I13.9\uparrow$. In the stuck-on piston fault scenario, the piston of the cylinder remained in the lower position after turning off the actuation for an indefinite time period. A possible hardware defect, which causes the piston to stuck-on, is a malfunctioning of the corresponding actuating valve.

Local FDI for $TAAO_3$
$FD(\widetilde{X}_{3,t}, c_3(t), \hat{\tau}_{up}(\widetilde{X}_{3,t})) = f_d$
$TRes_d^{\cap}(x_4) = \{I13.10\downarrow\}$
$TRes_d^{\cup}(x_4) = \{I13.10\downarrow\}$
Global FDI
$FD_{TDM} = f_d$
$TRes_{TDM,d}^{\cap} = \{I13.10\downarrow\}$
$TRes_{TDM,d}^{\cup} = \{I13.10\downarrow\}$

Table 5.2: Local and global FDI results for the stuck-on piston fault

The results of the online FDI are shown in Table 5.2. Given the last estimated state $\widetilde{x} = x_4$, no new behavior was observed until $c_3(t) = 81$ ms, hence a deadlock fault f_d was detected. The fault candidate, generated by global fault isolation, is $I13.10$. This I/O represents the limit switch that indicates the lower position of the piston. Since this sensor was mounted at the faulty cylinder, the fault related hardware was isolated by the fault candidate.

F2 Blocked Throttle Valve Fault

The next fault scenario considers the blocking of a pneumatic throttle valve that limits the speed of the transport unit assembled at station 3. The task of this transport unit was to collect workpieces coming from the rotary gripper of station 2, to deliver them to the presses and to forward them to the rotary gripper of station 4. The relevant behavior was modeled by the partial automata $TAAO_7$ and $TAAO_8$, depicted in Figures 5.17 and 5.18. Both partial automata modeled behavior of I/Os that belong to the transportation unit. Initially, the transportation carrier was located at the out-press. In order to receive a gear from the rotary gripper of station 2, the carrier had to be moved to the left hand position on its track. Therefore, the controller rose $O60.9\uparrow$ to open the actuating valve of the pneumatic system. After some time, the carrier was supposed to arrive at the left position. This was indicated by $I13.0\uparrow$, which led the controller to switch-off the moving actuator $O60.9\downarrow$ and to reset the stoppers $O60.7\downarrow$ and $O60.6\downarrow$. The stoppers are small cylinders mounted near the presses in order to stop the movement of the carrier at the designated positions. The air pressure of the carrier was limited by a throttle valve to reduce the speed of the carrier. If its speed was too high, the carrier hit the end limitations of the transportation track which caused mechanical damage after a number of incidences. In this fault scenario, the throttle valve of the pneumatic systems was blocked such that the pressure at the carrier was not reduced.

In Table 5.3, the results of the online FDI are shown. The local fault isolation had to be performed for the partial automata $TAAO_7$ and $TAAO_8$ since a fault f has been detected in both of them, simultaneously. Based on $TAAO_7$ and $\widetilde{x} = x_{28}$, a new partial system output $u_7(t)$ was observed at time $c_7(t) = 385$ ms leading to $vEdge(\lambda(x_{28}), u_7(t)) = \{I13.0\uparrow, O60.6\downarrow\}$. Since $u_7(t) = \lambda(x_{29})$ but $385 \notin [438, 1081]$, a fault f was detected.

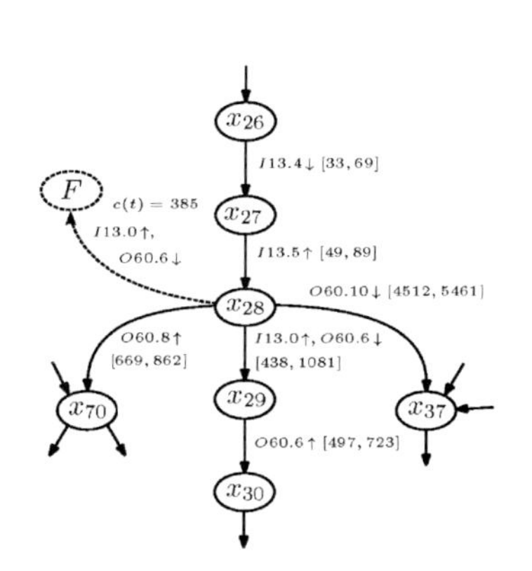

Figure 5.17: Detection of the blocked throttle valve fault via $TAAO_7$

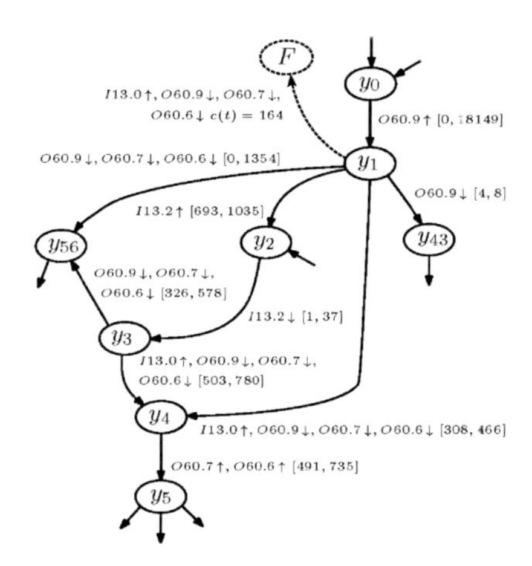

Figure 5.18: Detection of the blocked throttle valve fault via $TAAO_8$

Local FDI for $TAAO_7$
$FD(\widetilde{X}_{7,t}, c_7(t), \hat{\tau}_{up}(\widetilde{X}_{7,t})) = f$
$TRes_{eb}(x_{28}, u_7(t), 385) = \{I13.0\uparrow, O60.6\downarrow\}$
$TRes_{lb}(x_{28}, u_7(t), 385) = \emptyset$
Local FDI for $TAAO_8$
$FD(\widetilde{X}_{8,t}, c_8(t), \hat{\tau}_{up}(\widetilde{X}_{8,t})) = f$
$TRes_{eb}(y_1, u_8(t), 164) = \{I13.0\uparrow, O60.9\downarrow, O60.7\downarrow, O60.6\downarrow\}$
$TRes_{lb}(y_1, u_8(t), 164) = \emptyset$
Global FDI
$FD_{TDM} = f$
$TRes_{TDM,eb} = \{I13.0\uparrow, O60.9\downarrow, O60.7\downarrow, O60.6\downarrow\}$
$TRes_{TDM,lb} = \emptyset$

Table 5.3: Local and global FDI results for the blocked throttle valve fault

Considering $TAAO_8$, the last unambiguously estimated state before the detection of the fault was y_1. A new partial system output $u_8(t)$ observed at time $c_8(t) = 164$ ms can be reproduced by y_4 such that $u_8(t) = \lambda(y_4)$ but $164 \notin [308, 466]$. As a result, a fault f was detected based on $TAAO_8$ likewise. The global timed residuals determined the fault candidates $I13.0$, $O60.9$, $O60.7$ and $O60.6$. Since the failed actuating valve corresponds to $O60.9$, the fault can be exactly located by checking the provided candidates.

F3 Worn-out Test Cylinder Fault

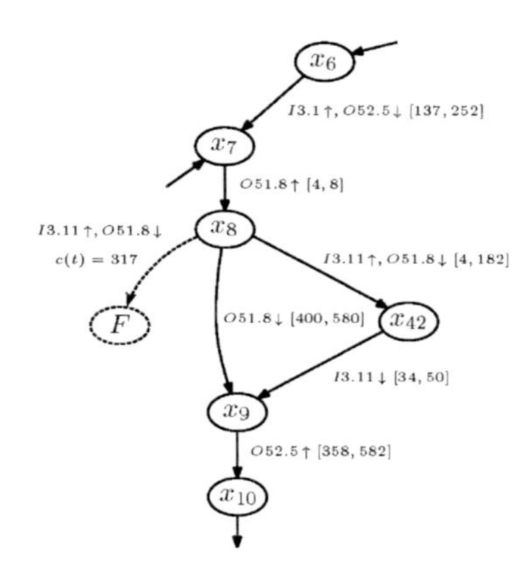

Figure 5.19: Detection of the worn-out test cylinder fault via $TAAO_2$

The third fault scenario is a worn-out test cylinder fault. The test cylinder is part of the sensor array located at station 2 of the BMS. Its task was to determine whether a gear coming from the input storage is equipped with a bearing. The behavior of the test cylinder was modeled by the $TAAO_2$, shown in Figure 5.19. During fault-free operation of the plant, gears were delivered to the position of the test cylinder indicated by $I3.1\uparrow$. After the position was reached, the transport unit was switched off $O52.5\downarrow$ and the test cylinder was extended, indicated by $O51.8\uparrow$. In the case that a gear had no bearing equipped, the cylinder was fully extended and the sensor $I3.11\uparrow$ changed its value. In case of an equipped bearing, the cylinder was not fully extended and $I3.11$ did not change its signal. Finally, the cylinder was retracted $O51.8\downarrow$ and the transportation unit was activated $O52.5\uparrow$ to forward the workpiece to the next sensor. In the fault scenario, the test cylinder was worn-out such that its extension and retraction speed was significantly reduced.

The behavior of the faulty BMS led the local FDI to detect a fault via $TAAO_2$. Given the unambiguous estimated state $\widetilde{x} = x_8$, the next partial system output $u_2(t)$ observed at $c_2(t) = 317$ ms led to the detection of f since $u_2(t) = \lambda(x_{42})$ but $317 \notin [4, 182]$. The local and the global residuals, given in Table 5.4, accordingly, led to the conclusion that

Local FDI for $TAAO_2$
$FD(\tilde{X}_{2,t}, c_2(t), \hat{\tau}_{up}(\tilde{X}_{2,t})) = f$
$TRes_{eb}(x_8, u_2(t), 317) = \emptyset$
$TRes_{lb}(x_8, u_2(t), 317) = \{I3.11\uparrow, O51.8\downarrow\}$
Global FDI decision
$FD_{TDM} = f$
$TRes_{TDM,eb} = \emptyset$
$TRes_{TDM,lb} = \{I3.11\uparrow, O51.8\downarrow\}$

Table 5.4: Local and global FDI results for the worn-out test cylinder fault

$I3.11$ and $O51.8$ are the fault candidates of the detected late behavior fault. Since both I/Os are related to the worn-out cylinder of the plant, the fault can again be located within the plant by the determined fault candidates.

The results obtained with the fault scenarios show that the proposed methods detect and isolate faults even when no information about the faulty system behavior is modeled. For system operators, it is most important to detect and isolate faults that make the production system fail. This happens most likely in the case of a deadlock and logical faults. However, based on the time fault symptoms early and late behavior, faults are detected and isolated that may currently not make the system fail but will most likely do so in nearer future.

5.6.3 Model Validation

Timed Identification

In Chapter 3, a timed identification approach for monolithic and distributed models was presented in which the time model behavior was determined by the proposed tolerance extension approach. Since fault detection capabilities of model-based FDI heavily depend on the applied FDI model, a validation with respect to the presented timed identification approach is given in the following. As discussed in Chapter 2, missed and false detections should be avoided during fault diagnosis of a DES. Therefore, two studies were made: In the first one, the number of time related false detections for different identified models was investigated and in the second one, it was checked whether an example set of time related faults could be detected.

The validation of false detections was based on a 10-fold cross validation principle. Therefore, the set of observed timed output sequences was divided into 10 parts of equal size without sharing sequences. A model was timed identified based on 9 of the 10 parts while the remaining one was used for validation. The final validation result was the mean value for all 10 validation runs. In that way, the validation can be considered as independent of the observed data. In order to investigate false detections of time faults, false detections resulting from logical non-reproducible behavior needed to be avoided. This was ensured by identifying the logical behavior of an automaton based on all observed

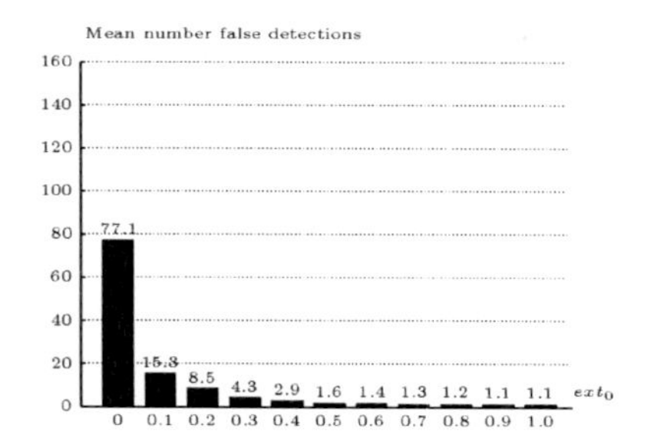

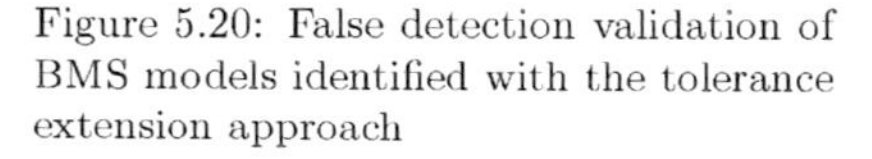
Figure 5.20: False detection validation of BMS models identified with the tolerance extension approach

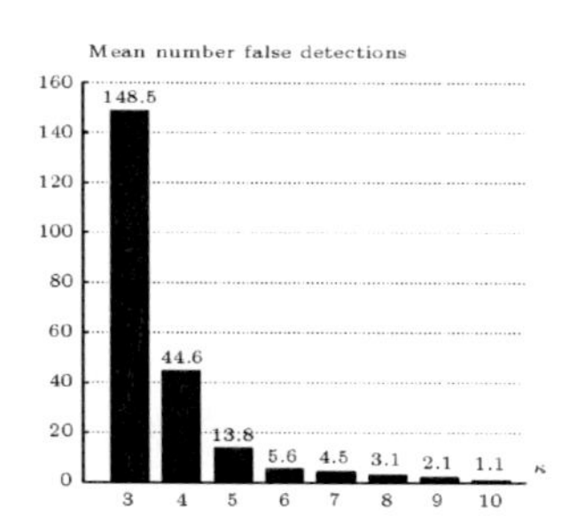

Figure 5.21: False detetion validation of BMS models identified with the normal distribution approach

sequences, independent of the sequence partitioning. Consequently, the models used for time validation were logical complete with respect to all observed sequences. The proposed tolerance extension approach for time guard determination was compared with the normal distribution approach for time guard determination after [Lefebvre and Leclercq, 2011]. In this approach, a time guard $[\tau_{lo}, \tau_{up}]$ is identified based on the estimated mean $\hat{m}$ and estimated standard deviation $\hat{s}$ for an observed transition time span sequence $\Delta_{Obs}^{(x,x')}$. The lower and upper time bounds are determined as $\tau_l = \max(0, \hat{m} - \kappa \cdot \hat{s})$ and $\tau_u = \hat{m} + \kappa \cdot \hat{s}$. The parameter κ is used to tune the width of the time guard interval in order to modify the probability that future observations lies within the identified time bounds. A detailed description of this approach is given in Chapter 6. The identification was performed using $|\Sigma_t| = 50$ observed timed sequences for a TDM with $N = 10$, including shared I/Os, the logical identification parameter $k = 2$, and the time identification parameters $1 - \alpha = 0.99$ and $\beta = 0.99$. The time identification parameters ext_0 and κ were varied, respectively.

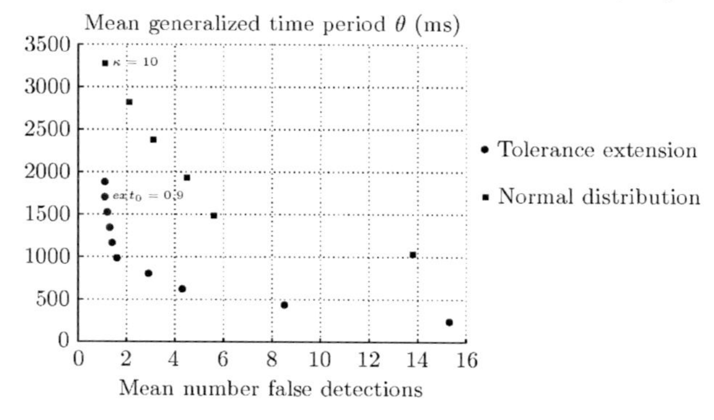

Figure 5.22: False detections and generalized time periods of timed identified BMS models

	Tolerance extension ($ext_0 = 0$)	Tolerance extension ($ext_0 = 0.9$)	Normal distribution ($\kappa = 10$)
F1 Stuck-on piston fault	detected	detected	detected
F2 Blocked throttle valve fault	detected	detected	not detected
F3 Worn-out test cylinder fault	detected	detected	detected

Table 5.5: Timed fault detection capabilities of timed identified BMS models

In Figures 5.20 and 5.21, the 10-fold cross validation results for the false detection experiments are shown. In total, 11 models based on the relative tolerance approach and 8 models based on the normal distribution approach were identified with different parametrization. The results show that increasing ext_0 and κ leads to models for which the number of fault detections decreases. This is because for higher values of ext_0 and κ, more tolerance is added to time guards which are observed for less than v_0. The higher tolerance raises the chance to obtain time guards that completely model the original time related system behavior. In that way, the non-reproducible time behavior of the models is reduced and less false detections have to be expected. However, reducing the non-reproducible time behavior typically comes along with an increase of the exceeding time behavior. A temporally precise model with minimum exceeding time behavior was identified by choosing $ext_0 = 0$. In this case, 77.1 false detections were made in average based on 5 validation sequences. If this number of false detections cannot be accepted, tolerance needs to be added to the identified time guards by a proper choice of ext_0 or κ. This tuning procedure, made by an engineer, aims to achieve an appropriate trade-off between the number of false detections and the probability of missed time faults that have to be expected during FDI.

In order to estimate the exceeding time behavior of the identified models, the generalized time period θ, introduced in Equation 3.48 in Chapter 3, is considered. θ represents the amount of time added to the minimum and maximum observed time spans for the determination of the time guards. The relation between the false detections and θ for the identified models is depicted in Figure 5.22. One can see that for a given number of false detections, the models identified with the relative tolerance extension approach contain less generalized time θ than models identified with the normal distribution approach. This means that models identified based on relative tolerance extension have a comparable higher time precision. For instance, consider the model identified based on the tolerance extension approach with $ext_0 = 0.9$ and the identified model based on the normal distribution approach with $\kappa = 10$. The mean number of false detections is 1.1 for both models while $\theta = 1705\,\text{ms}$ results for the tolerance extension approach and $\theta = 3272\,\text{ms}$ for the normal distribution approach. Hence, using the relative tolerance approach, less time needs to be added in order to achieve the same small number of false detections. This leads to a higher model precision and the chance of missed detections is comparable smaller. Consequently, the relative tolerance extension should be preferred for the time identification in this application.

The examination of the time related fault detection capabilities for different models was made upon the timed faults introduced in the preceding section. The results are given in Table 5.5. While the blocked throttle valve fault was detected with the models identified with relative tolerance extension, it could not be detected with a normal distributed model. The reason therefore is that the model identified with the normal distribution approach was too permissive with respect to the timed behavior of the relevant transition. This results from the fact that comparable larger generalized behavior was required to identify a model with a low false detection rate, as shown in Figure 5.22. The stuck-on piston fault and the worn-out test cylinder were detected with all investigated models.

Timed Distributed Modeling

In the following, the number of false detections and the fault detection capabilities are investigated for a monolithic model (M1), a distributed model (M2) partitioned by an expert, and a distributed model (M3) automatically partitioned according to the approach presented in Chapter 4. For the distributed expert model M2 and automatic partitioned model M3, $N = 10$ holds, respectively. All models were logically identified based on $k = 1$ and temporally identified using the tolerance extension approach based on the parameters $1 - \alpha = 0.99$, $\beta = 0.99$, and $ext_0 = 0.9$. Expert knowledge has only been used for the partitioning of the distributed expert model, not for its identification. The false detection validation was performed using again the 10-fold cross validation principle. In contrast to the time identification validation, in which the logical behavior of the model was identified based on the entire set of observed sequences, the training sets of the cross validation were used for both logical and time identification. In that way, logical and time related false detections could be studied.

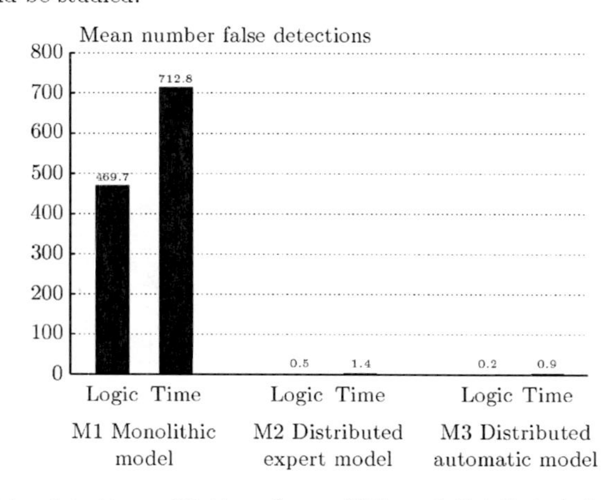

Figure 5.23: False detection validation of monolithic and distributed timed BMS models

In Figure 5.23, the results of the false detection validation for the three identified

	M1 Monolithic model	M2 Distributed expert model	M3 Distributed automatic model
F1 Stuck-on piston fault	not detected	detected	detected
F2 Blocked throttle valve fault	detected	detected	detected
F3 Worn-out test cylinder fault	not detected	not detected	detected
F4 Sensor short-circuit fault	not detected	detected	detected

Table 5.6: Fault detection capabilities using monolithic and distributed models

models is depicted. Considering the monolithic model, a large number of logical and time false detections was made during the 10 validation runs, based on 5 validation sequences, respectively. The major reason for this poor result is the concurrent behavior of the BMS. The set of observed timed output sequences used for identification did not represent the entire concurrent behavior of the system. Hence, the monolithic model was incompletely identified and many false detections were obtained during fault diagnosis. In contrast to that, fault detection based on distributed models led to significant better results. The distributed models were partitioned such that the related partial models could be approximately completely identified. Almost no false detections occurred and the number of false detections obtained with the automatically partitioned model is even slightly smaller than obtained with the expert partitioned one. This shows that the automatic partitioning approach determined a proper I/O-partition of the BMS, which could be completely identified based on the available set of observations.

To demonstrate the fault detection capabilities of the three FDI models, four BMS faults were investigated. In addition to the already introduced time faults F1, F2, and F3, the logical fault "F4 sensor short-circuit" was considered. The fault relevant sensor was located at the transportation unit of station 2. Its task was to recognize the arrival of the transporter carrier at the test cylinder in the middle of the station. When the carrier arrived at the cylinder, the sensor responded with a rising signal edge. When the carrier left the cylinder, a falling edge was observed. In the fault scenario, the sensor had a short-circuit leading to a rising edge at the related controller I/O although the carrier was not present at the sensor. This is an unexpected logical behavior. The fault detection results, given in Table 5.6, show that almost all faults were detected with the distributed models. The only fault missed by distributed FDI is the worn-out test cylinder fault F3 using the distributed expert model M2. This is because the relevant time guard was too permissive with respect to the late behavior of the cylinder. In contrast to that, all faults were detected with the distributed automatic model M3. The reason for the three missed detections made with the monolithic model refers to the high number of false detections. In Section 5.3, it has been explained that an unambiguous state estimation $|\widetilde{X}_t| = 1$ is a precondition to detect a fault. However, on each false detection, $\widetilde{X}_t = \emptyset$ results and fault detection can no longer be performed until the current model state $\widetilde{X}_t$ is again unambiguously estimated. Therefore, a number of observations is required. Due to the high number of false detections made with the monolithic model, the state estimator often entered the reinitialization phase and missing faults became highly probable.

Chapter 6

Related Works – Analysis and Comparison

6.1 Modeling of Timed DES

6.1.1 Language Model

In [Cassandras and Lafortune, 2008], three languages concepts are introduced to model the behavior of a DES on different abstraction levels: The *logical language*, the *timed language*, and the *stochastic timed language*. The language that models a DES in the most abstractive way is the logical language. It represents the logical behavior of the DES by considering the events and their ordering. The language that describes the DES behavior in a more detailed manner is the timed language. It models the time system behavior, in addition to the logical behavior, by adding information about the time instances when events occur. If statistical information about the system behavior is available, it can be used to model the stochastic timed language of the DES. The information can be, for instance, probability distributions about lifetimes of events or frequencies of event patterns. The stochastic timed language can be converted into the timed language by omitting the stochastic information. The timed language, again, can be condensed into the logical language by omitting the time information.

A well-known theory for timed languages of DES is proposed in [Alur and Dill, 1994]. The intended purpose of this language is to model the behavior of real-time systems over time. In [Alur and Dill, 1994], a *time sequence* $\tau = \tau_1\tau_2\ldots$ is introduced as the time instances of occurring events. It is an infinite sequence of time values $\tau_i \in \mathbb{R}$ with $\tau_i > 0$, satisfying the following constraints:

1. Monotonicity: τ increases strictly monotonically; i.e., $\tau_i < \tau_{i+1}$ for all $i \geq 1$.
2. Progress: For every $t \in \mathbb{R}$, there is some $i \geq 1$ such that $\tau_i > t$.

Since two events never occur at the same time instance, τ_i is monotonically increasing. The progress property ensures that the time sequences are basically infinite. For any time t, is exists at least one event that occurs later than t. Combining the time sequence with the related sequence of events leads to the definition of a *timed word*. A timed word over an alphabet E is a pair (σ, τ), $\sigma = e_1 e_2 \ldots$ is an infinite word over E, and τ is a time sequence. The logical word σ is an infinite sequence of symbols $e_i \in E$. Each unique event of the DES is considered as a unique symbol in the alphabet. With the timed sequence τ, a time attribute τ_i is unambiguously associated to each symbol e_i that represents the time when e_i occurs.

Example 6.1 (Infinite Timed Word).

$$(a, 2) \rightarrow (b, 2.7) \rightarrow (a, 3.3) \rightarrow (c, 4) \rightarrow \ldots$$

for a given alphabet $E = \{a, b, c\}$.

Definition 37 (Timed Language). "A timed language over E is a set of timed words over E" [Alur and Dill, 1994].

In contrast to the language definition given in [Alur and Dill, 1994], the authors of [Cassandras and Lafortune, 2008] define a language as a set of words such that each word has a finite-length n. The length n refers to the number of events from the event set E that are concatenated to a single word.

6.1.2 Automaton Model

A model that is able to represent a DES language is the automaton. Many definitions in literature exist differing in formalization and the language they can reproduce. To represent logical languages, appropriate input and output automata are available. Two examples in this context are Moor- and Mealy-automata [Cassandras and Lafortune, 2008]. In order to represent timed and stochastic timed languages, more complex model definitions are required. In the following, the *Timed Automaton with Guards*, according to [Cassandras and Lafortune, 2008], for modeling timed languages is introduced. It is defined as the six-tuple

$$G_{tg} = (X, E, C, Tra, Inv, x_0) \tag{6.1}$$

X is the set of states, E is the set of events, C is the finite set of clocks $c_1, \ldots, c_n$, with $c_i(t) \in \mathbb{R}^+$, $t \in \mathbb{R}^+$, Tra is the set of timed transitions of the automaton with

$$Tra \subseteq X \times \mathcal{C}(C) \times E \times 2^C \times X \tag{6.2}$$

$\mathcal{C}(C)$ is the set of admissible constraints for the clocks in set C, Inv is the set of state invariants, $Inv\colon X \rightarrow \mathcal{C}(C)$, and x_0 is the initial state. The time behavior modeled by the automaton is given by the clocks C, the admissible clock constraints $\mathcal{C}(C)$, and the state invariants Inv. If these elements of the automaton are removed, the definition of the model is reduced to a untimed automaton. The clocks C represent the time state of the model. Defining more than one clock allows to model multiple independently running, timed processes. The values of the clocks increase synchronously with the time t. A clock can be considered as a stopwatch that keeps on counting until it is reset at some time instance. The constraints that restrict the behavior of the model with respect to the time are the *admissible clock constraints* $\mathcal{C}(C)$. They are inequalities of the form

$$c_i(t) < q,\ c_i(t) \leq q,\ c_i(t) > q,\ c_i(t) \geq q \tag{6.3}$$

with q as a non-negative integer and c_i is the i-th clock of C, see [Cassandras and Lafortune, 2008]. The inequalities represent lower and upper bounds for the value of c_i. The *state invariants* Inv are admissible clock constraints that are assigned to states. In contrast to clock constraints, which are defined at transitions to passively restrict the timed state switching, the state invariants actively execute transitions. They force an automaton to leave an active state before the constraint, given by the invariant, is violated. This is especially useful in the case that a deadlock behavior due to violated clock constraints has to be avoided. A transition from state x to a following state x' is given as

$$(x, guard, e, reset, x') \in Tra \tag{6.4}$$

with $x, x' \in X$, $guard \in \mathcal{C}(C)$, $e \in E$, and $reset \subseteq C$. Upon a state transition the clock constraint is either satisfied $guard = true$ or violated $guard = false$ depending on the state of the clocks. The reset condition determines which of the clocks are reset upon the execution of the state transition. The admissible clock constrains $guard$ of a transition are denoted as *time guard* in the following.

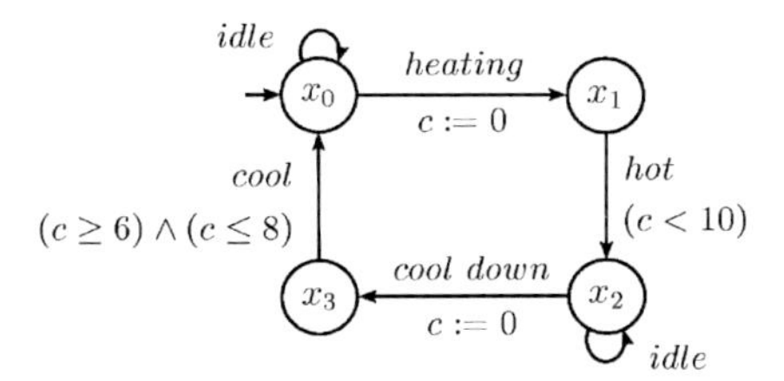

Figure 6.1: Example of an Timed Automaton with Guards

In Figure 6.1 an example of a Timed Automaton with Guards is given that models a heating and cooling process. The clock c monitors the time behavior of the process. One can see that time guards and resets conditions are not defined at all transitions. The modeled process starts in state x_0 upon the occurrence of the *heating* event. The transition from x_0 to x_1, which is associated with this event, has no time constraint but resets the clock $c := 0$. In state x_1, the automaton expects the event *hot* to occur before 10 time units after the activation of x_1. On its occurrence, an arbitrary long *idle* phase is modeled that ends with the occurrence of the *cool down* event. The automaton can stay in state x_2 an arbitrary long time since there exists at least one transition that has no time constraint. When the *cool down* event occurs, clock c is reset again and the next event *cool* is expected to occur within 6 and 8 time units. This means that state x_3 is active for at least 6 and at most 8 time units. The occurrence of the *cool* event leads the system to return to its initial state x_0.

6.1.3 Petri Net Model

The second model class that is commonly used to represent languages of DES are Petri nets. As well as for automata, many different formalisms for Petri nets have been introduced in literature. To model timed languages, the *timed Petri net (TPN)* is an appropriate representative. In the following, the definition according to [Berthomieu and Diaz, 1991] is introduced as the six-tuple

$$\mathcal{N} = (P, T, B, F, Mo, SIM) \tag{6.5}$$

P is a finite nonempty set of places p_i, T is a finite nonempty set of transitions t_i, B is the backward incidence function $B\colon T \times P \to N$ with the set of non-negative integers N, F is the forward incidence function $F\colon T \times P \to N$, Mo is the initial marking function $Mo\colon P \to N$, and SIM is a mapping called static interval

$$SIM\colon T \to Q^* \times (Q^* \cup \infty) \tag{6.6}$$

with the set of positive rational numbers Q^*. The places, transitions and the incidence functions describe the logical structure of the model. Incidence functions connect the transition and places and add weights in form of non-negative integers to the resulting arcs. The time behavior is represented by the static intervals that are assigned to each transition. They correspond to the time guards that are defined for the timed automaton. In [Berthomieu and Diaz, 1991], a distinction between *static* and *dynamic intervals* is drawn. The static interval for a transition t_i is given as $SIM(t_i) = (\alpha_i^S, \beta_i^S)$ with α_i^S as lower bound and β_i^S as upper bound, respectively. While the static intervals remain constant during model execution, the dynamic intervals (α_i, β_i) change with the progress of time. The lower bound α_i is called earliest firing time (EFT) and the upper bound β_i is called latest firing time (LFT) of transition t_i.

To discuss the dynamic properties of a TPN the notation of state is introduced. A state S is defined as a pair $S = (M, I)$, with marking M and firing interval set I. The marking M represents the number of tokens assigned to each place while the firing interval set I contains the intervals of those transitions that can fire based on the given marking. A transition t_i of the TPN is allowed to fire at time $\tau_{Abs} + \theta$ according to [Berthomieu and Diaz, 1991] if and only if the two following conditions are fulfilled:

1. t_i is enabled by marking M at time τ_{Abs}:

$$(\forall p)(M(p) \geq B(t_i, p)); \tag{6.7}$$

2. the relative firing time θ is not smaller than the EFT of transition t_i and not greater than the smallest LFT of all the transitions enabled by marking M:

$$\text{EFT of } t_i \leq \theta \leq \min\{\text{LFT of } t_k\} \tag{6.8}$$

with k denoting the transitions that are enabled by M. The first condition expresses the basic enabling condition that is the same for untimed Petri nets. A transition t_i is enabled if the number of tokens in all preceding places p of t_i is at least as large as the weight of the arcs. The second condition evaluates the time bounds of the transition. The transition t_i can fire if a relative firing time θ expires that is larger than the EFT of t_i and smaller than the minimal LFT of all transitions t_k enabled by marking M. The relative firing time θ is thereby considered with respect to the absolute time τ_{Abs} and can be seen as an expected waiting time. If a transition is allowed to fire, the state of the TPN according to [Berthomieu and Diaz, 1991] is changed as follows:

$$(\forall p)M'(p) = M(p) - B(t_i, p) + F(t_i, p) \tag{6.9}$$

and I' is derived by taking the updated intervals of remaining active transitions and adding the static intervals of new enabled transitions. After firing a transition, tokens are removed from the preceding places and added to the following places. The number of tokens removed or added corresponds to the weights of the arcs, respectively. The new firing interval set I' is constructed by updating the dynamic intervals of the remaining active transitions with the relative time passed θ. If new transitions are enabled after firing, their static intervals are added to I'.

For the representation of timed languages with the introduced TPN, an additional function is necessary to incorporate the events of the DES, namely the labeling function $\lambda\colon T \rightarrow E$. The function assigns an event of the event set E to each transition. Petri nets of that type are often called *labeled Petri nets* in literature [Cassandras and Lafortune, 2008].

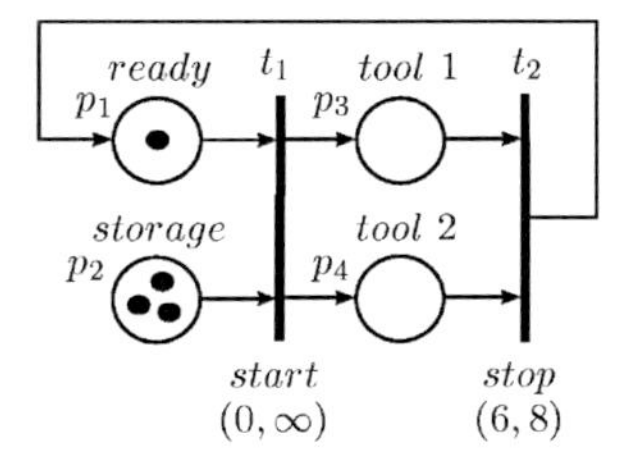

Figure 6.2: Example timed Petri net with labels

An example of a TPN with labels is given in Figure 6.2. It models the time bounded operation of two different tools that simultaneously process work pieces from a storage. At the beginning, place p_1 has one marking, p_2 has 3 markings, and the other places are unmarked. In this situation, transition t_1 is allowed to fire upon the occurrence of the event *start* within the maximum permissive firing interval $(0, \infty)$. When the transition fires, a marking is removed from p_1 and p_2 and added to p_3 and p_4, respectively. The weights of all arcs in this example are equal to one. After that, transition t_2 is allowed to fire within the time interval $(6, 8)$. When the *stop* event is observed in time, p_3 and p_4 are unmarked and one marking is added to p_1. The tools are again ready for processing while only two items are left in the storage, indicated by the two remaining markings. The process can be executed as long as there exist markings in p_2.

6.1.4 Discussion

The timed language of [Alur and Dill, 1994] is an appropriate concept to formally model the timed behavior of a DES. Many different models are proposed in literature that are able to represent timed languages in a compact way. Besides timed automata and Petri nets, two other modeling formalisms are *neural networks* and *dioid algebra*. Neural networks are structures of interconnected neurons that can be trained to model the logical and time behavior of a system [Choi and Kim, 2002]. These structures are by nature not interpretable and cannot explicitly be used to derive information for fault isolation. Another modeling approach for representing the timed system behavior of a DES is the dioid algebra [Cassandras and Lafortune, 2008]. This algebra defines the two atomic operations "max, +" or "min, -" that are used to formulate specific state equations. The framework provides numerous analytical tools but modeling becomes very complicated for large systems and model identification is hardly possible.

The two modeling frameworks that are commonly used in the context of identification and timed fault diagnosis of DES are *timed automata* and *timed Petri nets*. Their advantage is, compared to many other modeling concepts, the explicit representation of the system behavior in terms of states and transitions. Automatic modeling is well studied in literature for both with respect to many problems and applications. Although automata and Petri nets are closely related, they have some individual strengths. An advantage of automata is the intuitive representation of state in the sense that one system state corresponds to one model state [Cassandras and Lafortune, 2008]. They allow for component wise modeling and the building of global models by composition operations. In general, more analytical tools come along with automata than with Petri nets. A disadvantage of automata is the state-space explosion problem that can arise for large systems. However, this problem can be avoided by using appropriate model structures. The advantage of Petri nets is that it can represent more languages than automata using the same memory space. A automaton can always be transformed into a Petri net but the other direction is not always possible if the state-space of the automaton should be finite [Cassandras and Lafortune, 2008]. Petri nets are somehow a more general modeling formalism. Another advantage is the modeling of concurrency in a compact representation. On the other hand, Petri nets have a less intuitive representation since the information about the system state can be distributed among several marked places [Cassandras and Lafortune, 2008]. This holds especially for large nets with multiple marked states. Another disadvantage is that component wise modeling is not supported which makes modeling of large systems in general difficult. In summary, the choice of which model is the preferably used depends on the system that is to be modeled, the required tools, and finally also on the engineers personal preferences.

6.2 Identification of DES Models

6.2.1 Preliminaries

The traditional way of modeling is to formulate the relevant physical relationships of a system in terms of algebraic equations. This requires deep knowledge about the physical conditions, constraints, and the dynamics of the system. If this knowledge is not available because the system is large, complex, or hard to access, then *identification* methods are applied.

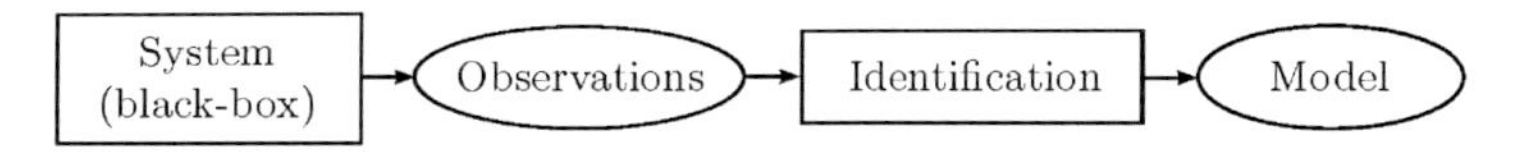

Figure 6.3: Principle of passive model identification

Model identification is originally a branch of artificial intelligence in the field of computer science. The idea is to observe the input and output behavior of a system in order to build a model that is able to reproduce its behavior. Since the structure and dynamics of the considered system are unknown, it is referred to as a *black-box*. Identification approaches can be classified according to several criteria. With respect to their operation principle, identification can be either performed *actively* or *passively*. Active identification algorithms excite an unknown system according to predefined test sequences and observe its reactions. The test sequences are typically defined in a way such that the entire behavior of the system can be observed. In contrast to that, passive identification algorithms rely on data that is obtained by passive observation of the system. This data can be, for instance, recorded during normal operation of the system. Passive identification does not affect the DES and cannot influence the kind of observed data. However, if a DES is observed for a sufficient long time, the obtained observations likely contain the information about the complete system behavior. Figure 6.3 illustrates the passive identification principle. Another characteristic of identification algorithms concerns the type of observation data that is used. *Positive observations* represent behavior the system can actually perform. Typically, they can be observed during the system's operation. Counterexamples of the original system behavior are denoted as *negative observations*. This behavior cannot be performed by the system, naturally.

Several decades ago, [Gold, 1978] proposed the idea of "identification in the limit" as a general theory for identification. Based on an infinite sequence of observations of an unknown system, the task of identification is to find a model such that it can be considered as completely identified in the limit. This means, after a finite number of observations D_i a guessed model $g(D_i)$ can be determined that is able to represent the behavior $g(D_{i+1})$ for any following growing data set D_{i+1}. It is assumed that the behavior of the model *converges* to an unknown limit. This assumption is fundamental for identification approaches since it ensures that it is in general possible to completely identify models. To perform system identification, a basic model class and a convergence criterion must be defined [van Schuppen, 2004]. The model class has to be chosen in a way that the relevant phenomena are modeled and the identification problem is analytically, algebraically, and computationally treatable. Models that fulfill these requirements have been introduced in the preceding section. They explicitly model the timed behavior and can be computationally handled due to their well-defined formalization. Based on the observations of the system, the selected model is identified with respect to a defined quantitative convergence criterion. The criterion is necessary to determine the achieved modeling accuracy and to compare different model designs.

In literature, identification approaches for different modeling formalisms are proposed. They can be subdivided into approaches that identify *logical* and *timed models*. While logical identification aims to build models that represent the observed events and their logic sequence of occurrence, the timed identification approaches aim to generate models that additionally represent the event timings. Since this thesis focuses on the generation of timed models, most of the following presented methods fall into this category. Nevertheless, in order to give a complete overview about the identification approaches for DES, the identification of logical models is also briefly discussed in the following.

6.2.2 Identification of Logical Models

The early works about automata identification were done in the 1960s. It started with learning of stochastic automata for the modeling of stochastic processes. The identification task was to adapt the state transition probabilities according to observations that were made, see [Narendra and Thathachar, 1974]. In the following decades, numerous approaches for the identification of models that represent the logical behavior of a system were presented by the scientific community. In the work of [Booth, 1967], an active approach for the identification of Moore- and Mealy-automata is described. These models are a special class of input and output automata, see [Cassandras and Lafortune, 2008]. To generate the necessary data for identification, the inputs of a system are stimulated by predefined input sequences and the resulting output sequences are observed. The obtained sequences are used to identify an automaton that describes the complete behavior of the system under investigation. A logical identification approach for automata, which relies on data from passive observations, is given in [Veelenturf, 1978]. The identification algorithm builds the automaton in a step wise manner. It adds states and transitions to the model for each observation until a model is obtained that completely represents the observation data. In [Klein, 2005] and [Roth, 2010], the problem of model identification for the purpose of fault diagnosis in closed-loop DES is addressed. The authors show that the automatic identification approaches proposed in the scientific community so far do not generate appropriate models for fault diagnosis purposes. They proposed a new identification algorithm that builds a *Non-Deterministic Autonomous Automaton with Outputs* (NDAAO) using passively obtained input and output observations. This model is used to reproduce the fault-free behavior of a DES in order to perform fault detection and isolation. Given an identification parameter k, the identification algorithm uses observed words of different length to build the state-space, the transitions, and the output set. An output is mapped to each state such that state trajectories represent the event sequences of the DES. In contrast to other identification approaches, precision and completeness guarantees are given for the identified logical models. The mentioned identification approaches so far do all rely on positive samples of the system behavior. A comprehensive overview about of other automata identification approaches for logical models is given in [Klein, 2005].

As for automata, several approaches for the identification of logical Petri net models exist in literature. In [Meda-Campaña and López-Mellado, 2005], the *stepwise identification* of a labeled Petri net models is studied. Similar to the approach of [Veelenturf, 1978], a Petri net is built by successively adding places and transitions with each new observation. The observed data are passively obtained during normal operation of the system. It is assumed that after a sufficient number of observations the model asymptotically converges to the complete model of the system. This assumption is consistent with the identification in the limit assumption proposed by [Gold, 1978]. In the work of [Cabasino et al., 2007], another approach for the identification of labeled Petri nets is presented. The identification problem is formulated in terms of an *integer linear programming problem (ILP)*. The task is to synthesize a Petri net N with initial marking M_0 such that its language $L^n(N, M_0)$ is equal to the unknown language L with n as the length of the longest string in L. A net system $\langle N, M_0 \rangle$ is a solution of this optimization problem if and only if it is able to represent positive observations and unable to represent negative observations of the system behavior. If the dynamics of the system are slow with respect to the time needed to solve the ILP, then the identification approach can be applied in real-time. A similar approach is given in [Dotoli et al., 2008] in which only positive observations are applied for identification. For a detailed overview about identification approaches for logical Petri nets, the reader is referred to [Fanti and Seatzu, 2008].

6.2.3 Identification of Timed Models

The active identification of timed automata is considered in [Grinchtein et al., 2005]. Given an unknown timed automaton A, the identification task is to build a hypothesized timed automaton H that accepts the same timed language as the unknown one $L(A) = L(H)$. The identification is carried out by a *timed learner* and a *timed teacher*. The timed learner has no initial knowledge about A and asks queries to the teacher in order to learn $L(A)$. The teacher is aware of A and answers the queries of the learner. Two different types of queries exist: *membership queries* and *equivalence queries*. A membership query is to ask whether a timed word w, guessed by the learner, is in the language of the system $w \in L(A)$. The response of the teacher is either "yes" or "no". Obviously, the learner has to know about basic alphabet of the unknown system in order to ask these queries. After asking a collection of membership queries, the learner constructs the identified automaton H that satisfies the answered queries. He submits H in an equivalence query to the teacher and asks whether its language is equivalent to the language of the unknown automaton $L(A) = L(H)$. If the language equivalence is confirmed by the teacher, the learning procedure is complete, if it is denied, the teacher provides a counterexample that is in $L(A) \setminus L(H)$ or in $L(H) \setminus L(A)$. The counterexamples are used to extend the set of membership queries in order to refine the identified model.

An example for the identification of an timed automaton according to [Grinchtein

Membership query	Response
λ	yes
$(a, x_a = \perp)$	no
$(a, x_a = \perp)(a, x_a = 0)$	yes
$(a, x_a = \perp)(a, 0 < x_a < 1)$	no
$(a, x_a = \perp)(a, x_a = 1)$	yes
$(a, x_a = \perp)(a, x_a > 1)$	yes

Table 6.1: Answered membership queries after the second identification step

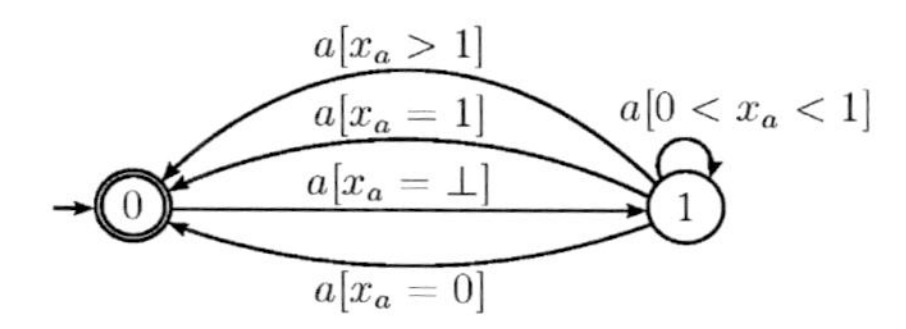

Figure 6.4: Identified automaton H after the second identification step [Grinchtein et al., 2005]

et al., 2005] is given in Table 6.1 and Figure 6.4. The table and the figure represent the intermediate results after the second identification step. The membership queries asked by the learner in the first step are the empty string λ and the time guarded word $(a, x_a = \perp)$, with event $a \in E$ and x_a is the relative time elapsed since the last occurrence of a. Since there is no preceding occurrence of a, the value of clock x_a is undefined, denoted by $\perp$. The response of the timed teacher shows that the empty set is accepted by the unknown automaton while $(a, x_a = \perp)$ is not. After the first identification step no automaton is built since the data is not sufficient. Four more time guarded words are asked in the second step by the learner and finally the hypothesized automaton in Figure 6.4 is constructed. A time guarded word that is accepted by the teacher starts and ends in the marked state indicated by the doubled circle. The automaton is then sent to the teacher for the equivalence check. In this example, the teacher denies the equivalence and provides the counterexample $(a, x_a = \perp)(a, x_a = 0)(a, x_a = 0)(a, x_a = 0)$. This counterexample is used to generate new membership queries in order to update the model until it is complete.

In [Verwer et al., 2006], an identification approach for timed automata is proposed that is based on a *state merging* algorithm. As in [Grinchtein et al., 2005], the model that is to be identified is a timed automaton with guards. The data used are positive and negative observations of the system behavior. The aim of the identification is to construct a timed automaton with minimum number of states that is consistent with the observations. The algorithm starts with the construction of the timed *Prefix Tree Acceptor (PTA)*. A timed PTA is a tree-like automaton whose branches represent either a positive or a negative observed sequence. Since timed sequences are typically all different from each other due to the time variations, most of the branches represent a single observed sequence. Positive and negative behavior is distinguished in the PTA by different markings of the final states. After construction of the timed PTA, the state merging algorithm is applied. It merges equivalent states q_1 and q_2 into a new state q' and reconnects the transitions such that the automaton still consistent with the observations. Transitions with identical symbols are merged into one as well as their time guards and target states. This done according two rules. The first rule is to merge two transitions if they have the same symbol and the same target state. The time guard of the resulting transition is $[\min(l_1, l_2), \max(h_1, h_2)]$ with $[l_1, h_1]$ and $[l_2, h_2]$ representing the time guards of the original transitions, respectively. The second rule is to merge transitions if they have the same symbol and overlapping time guards. The guards are combined in the same way as with the first rule.

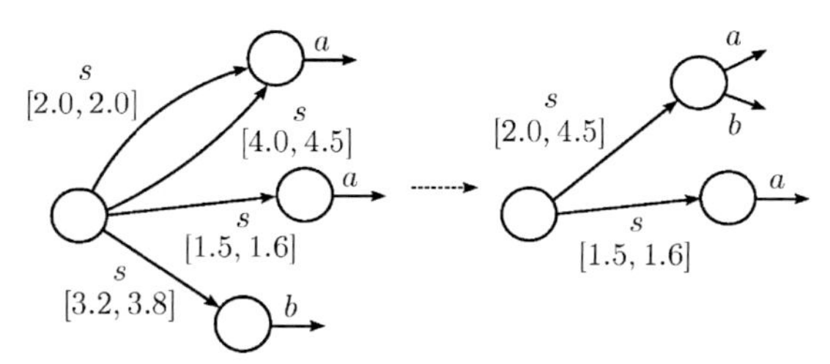

Figure 6.5: State merging in the prefix tree acceptor

In Figure 6.5 an example for the outlined state merging rules is given. In order to reduce the number of states of the automaton on the left hand side, two consecutive state merging operations are performed. According to first rule, the transitions $(s, [2.0, 2.0])$ and $(s, [4.0, 4.5])$ are merged to $(s, [2.0, 4.5])$ since they have the same event s and the same target state. Next, the resulted transition $(s, [2.0, 4.5])$ is merged with $(s, [3.2, 3.8])$ since they have the same event s and their time guards are overlapping. This merge is performed according to the second rule. As a result, the transition $(s, [2.0, 4.5])$ and the merged target state with outgoing events a and b are obtained. The two remaining transitions cannot be further merged since they have neither the same target state nor overlapping time guards.

An identification approach developed in the context of fault diagnosis is proposed by [Supavatanakul et al., 2006]. The work addresses the identification of a non-deterministic timed automaton with guards that explicitly models the fault-free and faulty input and output behavior. In the work, systems with continuous I/O-signals are considered. These signals are quantized such that the system generates discrete events. An event is raised if the value of the measured signal leaves a quantization interval and enters another one. This is indicated by one or more changing I/Os. In order to represent this behavior by the model, the timed automaton contains input and output conditions that are associated to the transitions. They define the logical behavior of the model. The time behavior of the automaton is determined by relative time intervals $[\tau_{min}, \tau_{max}]$, with τ_{min} and τ_{max} denoting the earliest and latest time the transitions can occur, respectively.

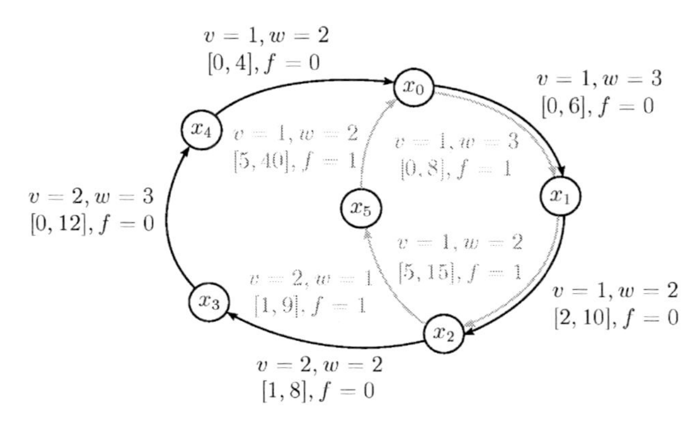

Figure 6.6: Non-deterministic timed automaton with time guards

In Figure 6.6, a timed automaton is shown that models the fault-free and faulty behavior of a DES. The transitions with the fault label $f = 0$ generate events that correspond to the fault-free behavior of the system. The events based on a faulty DES behavior refer to the gray transitions $f = 1$. Different fault labels can be used to indicate different types of faults. The figure also illustrates the non-determinism of the automaton. For instance, the states x_0 and x_1 are connected by two different transitions. Both transitions have the same I/O signals ($v = 1, w = 2$) and share the relative time interval $[0, 6]$ in which both transitions can fire. However, since the transitions have different fault labels, one transition corresponds to the fault-free behavior and one to the faulty behavior of the DES, with subject to fault $f = 1$. If the signals ($v = 1, w = 2$) are observed at relative time $t = 3$, then the transition that fires is chosen in a non-deterministic way. The identification algorithm proposed by [Supavatanakul et al., 2006] uses observed timed I/O-sequences to build the timed automaton. The applied sequences represent the fault-free DES behavior $f = 0$ and the faulty behavior $f = i$ for different fault types $i = 1, 2, \ldots, S$. It is assumed that sufficient data for the identification of the fault-free and faulty behavior is available. Initially, the state-space of the automaton is built by adding a state for each distinct observed event, i.e. for each observed combination of the I/O signals. Then, the transitions of the automaton are determined according to the following two rules. If a transition exists that is able to reproduce the logical behavior of an observation, only the time guard $[\tau_{min}, \tau_{max}]$ is updated such that $\tau \in [\tau_{min}, \tau_{max}]$ holds, with τ representing the relative time of the new observation. If the automaton cannot reproduce the observed logical behavior, a new transitions is added with $\tau_{min} = \tau_{max} = \tau$ and with the fault label $f = i$ depending on the applied observed sequence.

In the outlined identification approach of [Supavatanakul et al., 2006], the time guards are determined by the minimum and maximum time an event is observed. In contrast to that, the following approach applies a *statistical analysis* of the event timings to determine the time guards. The work, proposed by [Lefebvre and Leclercq, 2011], focuses on the identification of timed Petri nets for fault diagnosis of DES. The identification is performed in two steps. In the first step, the logical behavior of the Petri net is identified. This is done by generating a transition t_i for each event e_i, $i = 1, 2, \ldots, Q$ and connecting these transitions with Q places such that the resulting net is able to reproduce the observed event sequences. The sets of places and transitions are subsequently simplified by merging algorithms and the initial markings are derived. Once the logical structure of the TPN is build, the time properties are determined in the second step. For each transition, a time guard is identified according to the event timings. It is assumed that the timings are observed at randomly time instances according to *probability distributions*. While the type of distribution is assumed to be a priori known, the parameters have to be identified. The event timings that result from the fault-free system behavior are normally distributed while the timings of fault events are exponentially distributed. For a given event e_j, the parameters of a normal probability distribution function (pdf) are $\theta_j = (m_j, s_j)$, with m_j denoting the mean value and s_j is the standard deviation. The exponential pdf is characterized by the parameter $\theta_j = \mu_j = \frac{1}{m_j}$, with the firing rate μ_j. The aim of the time identification is to determine the unknown pdf parameters θ_j for each event e_j and to deduce proper time guards based on this information. For a given state x_i and a given event e_j, the pdf parameters $\theta_{i,j}$ are estimated using the following equations from [Lefebvre and Leclercq, 2011]:

$$\hat{m}_{i,j} = \frac{1}{n_{i,j}} \sum_{k \in E_{i,j}} (t(k) - t(k-1)) \tag{6.10}$$

and

$$\hat{s}_{i,j} = \sqrt{\frac{1}{n_{i,j}} \sum_{k \in E_{i,j}} ((t(k) - t(k-1)) - \hat{m}_{i,j})^2} \tag{6.11}$$

$\hat{m}_{i,j}$ is the estimated mean value, $\hat{s}_{i,j}$ is the estimated standard deviation, $E_{i,j}$ the set of discrete time steps k when $e(k) = e_j$ and $x(k) = s_j$, and $n_{i,j} = |E_{i,j}|$. The equations are applied to events that have no concurrent behavior with any other events of the DES. In case of concurrency, the estimated pdf parameters, based on these equations, may not represent the true parameters. Then, an additional iterative correction algorithm has to be applied. The correction algorithm basically compares the observed with the modeled timings that are produced by the Petri net on its execution. If there is a deviation, the parameters of the pfds are corrected accordingly. After estimating the pdf parameters, time guards for the normally distributed fault-free transitions are determined. They represent the fault-free time behavior of the DES. The resulting timed Petri net is used for fault diagnosis. No time guards are determined for the fault related transitions with exponential distributed pdfs. Instead, the parameters of the pdfs that represent a faulty behavior are used to isolate the detected faults. The time guards for the transitions with normally distributed timings are determined according to confidence intervals TI_j. These intervals are defined around the estimated mean value $\hat{m}_j$ with respect to a predefined risk level $1 - TF_j$. The equations are given in [Lefebvre and Leclercq, 2011] as:

$$TI_j = [\max(0, \hat{m}_j - \kappa_j \cdot \hat{s}_j), \hat{m}_j + \kappa_j \cdot \hat{s}_j] \tag{6.12}$$

and

$$TF_j = \int_{\hat{m}_j - \kappa_j \cdot \hat{s}_j}^{\hat{m}_j + \kappa_j \cdot \hat{s}_j} \left(\frac{1}{\hat{s}_j \cdot \sqrt{2\pi}} \cdot e^{-\left(\frac{(t-\hat{m}_j)^2}{2 \cdot \hat{s}_j^2}\right)} \right) \cdot dt \tag{6.13}$$

with the extension variable κ_j, whose value depends on $\hat{m}_j$, $\hat{s}_j$ and the choice of $1 - TF_j$.

To illustrate the identification of time guards, the following simple example is considered. It is assumed that a Petri net is logically identified based on data of the fault-free DES behavior. Further it is assumed that two places P_1 and P_2 exist that are connected by a single transition T_2. The event e_2 that belongs to T_2 has no concurrent behavior to any other event. The estimated normal pdf parameters that correspond to T_2 are $\hat{m}_2 = 5$ and $\hat{s}_2 = 2$. Given a risk level $1 - TF_2 = 0,26\%$, the parameter of the confidence interval is determined as $\kappa_2 = 3$, according to Equation 6.13. The confidence interval results in $TI_2 = [\max(0, 5 - 3 \cdot 2), 5 + 3 \cdot 2]$. If the parameters of the normal pdf are correctly estimated, the probability is 99,74% that an observed timing of event e_2 lies withing TI_2. Increasing this probability by a smaller choice $1 - TF$ would lead to a guard with wider limits.

Another Petri net approach that uses time information for the identification is proposed in [Basile et al., 2011]. Instead of generating explicit time models, the idea is to use time information in order to accelerate the logical identification procedure. The identification algorithm is based on the ILP approach proposed by [Cabasino et al., 2007], in which positive observations E and counterexamples D of the DES behavior are used to build a Petri net. It is assumed in the work of [Basile et al., 2011] that, besides the number of places and transitions, also the timings $\delta(t)$ for all transition t are known. The timing $\delta(t)$ is given as a time that must elapse before a transition t fires. This definition corresponds to a time guard in which the lower bound is equal to the timing $\delta(t)$ and the upper bound is unrestricted. The timing information is used by the identification algorithm to generate counterexamples of the system behavior. This helps to reduce the number of necessary optimization steps by excluding solutions of Petri nets that are not consistent with the required timings.

In [Das and Holloway, 2000] an approach is proposed that determines time guards for fault diagnosis models based on estimated confident spaces. The method does not rely on a specific model definition but can rather be applied to a class of models that use the time guard concept. The guards are identified based on observed timings of the fault-free DES operation such that the mismatch between the unknown true timing interval $\underline{T}_R$ and the determined time guards $\underline{T}_C$ is minimal. Both time intervals are represented by pairs $\underline{T}_i = (m_i, w_i)$, with m_i denoting the middle and w_i denoting the width of the interval. The lower and upper bound of an interval are given as $m - \frac{w}{2}$ and $m + \frac{w}{2}$, respectively. The true timing interval $\underline{T}_R = (m_R, w_R)$ represents the timings between two events that can occur during the fault-free operation of the DES. The exact values of m_R and w_R are unknown. However, it is assumed that a confident range is available for each parameter that contains the true value. In this work, these ranges are determined by assuming normal distributed timings and creating confidence intervals around the estimated normal pdf parameters. The two-dimensional space D, spanned by the confidence intervals for m_R and w_R, is called *confident space*. It contains the true timing interval $\underline{T}_R \in D$. Given the confident space D, the task is to determine an interval $\underline{T}_C$ such that the number of missed and false detections with fault diagnosis is minimal for any $\underline{T}_R$ in D. The measure to determine the normalized missed detection rate for a given $\underline{T}_C$ is introduced in [Das and Holloway, 2000] as:

$$MD(\underline{T}_R) = \frac{|\underline{T}_C| - |\underline{T}_R \cap \underline{T}_C|}{|\underline{T}_C|} \tag{6.14}$$

with $0 \leq MD(\underline{T}_R) \leq 1$. If $\underline{T}_C$ is completely covered by $\underline{T}_R$, then $\underline{T}_R \cap \underline{T}_C = \underline{T}_C$ and $MD(\underline{T}_R) = 0$. The more time instances of $\underline{T}_C$ lie outside of $\underline{T}_R$, the larger is $MD(\underline{T}_R)$. Respectively, the measure for the normalized false detection rate of a given $\underline{T}_C$ is introduced in [Das and Holloway, 2000] as:

$$FA(\underline{T}_R) = \frac{|\underline{T}_R| - |\underline{T}_R \cap \underline{T}_C|}{|\underline{T}_R|} \tag{6.15}$$

with $0 \leq FA(\underline{T}_R) \leq 1$. If $\underline{T}_R$ is completely covered by $\underline{T}_C$, then $\underline{T}_R \cap \underline{T}_C = \underline{T}_R$ and $FA(\underline{T}_R) = 0$. Timings that lie within $\underline{T}_R$ but outside of $\underline{T}_C$ lead to a false detection rate that is different from zero. In order to rate the choice of $\underline{T}_C$ with respect to the missed detection and false detection rate, the following cost function is introduced:

$$h(\underline{T}_R, \underline{T}_C) = K(FA(\underline{T}_R)) + (1 - K)(MD(\underline{T}_R)), \tag{6.16}$$

with the weighting factor $K \in [0, 1]$ that is used to make a trade-off between missed and false detections [Das and Holloway, 2000]. Since the confident space D contains a set of possible timings rather than only one, a time guard $\underline{T}_C$ has to be rated with respect to all possible time intervals $\underline{T}_R$ in D. In order to obtain a time guard that is valid for any $\underline{T}_R \in D$, the *worst case costs* have to be considered. These are the highest costs that are obtained based on D and a selection of $\underline{T}_C$. Finally, the aim is to find a $\underline{T}_C$ for a given K such that the worst case costs based on D are minimal. The equation is given as:

$$\underline{T}_C^* = \arg \min_{\underline{T}_C \in U} \left(\max_{\underline{T}_R \in D} \left(h\left(\underline{T}_R, \underline{T}_C\right)\right)\right) \tag{6.17}$$

with the set of all possible intervals U.

In the following, an illustrative example is presented that explains the optimal choice of a time guard $\underline{T}_C$. It is assumed that the confident space D is given as

$$D = \{(m, w) \mid (4 \leq m \leq 8) \wedge (2 \leq w \leq 4)\}, \tag{6.18}$$

the weighting factor is $K = 0.5$, the middle of $\underline{T}_C$ is $m_C = 6$, and the worst case costs are obtained with the intervals $\underline{A} = (8, 2)$ and $\underline{B} = (8, 4)$, $\underline{A}, \underline{B} \in D$. The value of m_C and the intervals $\underline{A}, \underline{B}$ are determined according to approach proposed in [Das and Holloway, 2000]. This will not be further discussed here. The confident space D is depicted in Figure 6.7. The aim is to determine a time guard $\underline{T}_C$ with middle $m_C = 6$ and width w_C such that the worst case costs obtained with $\underline{A}$ and $\underline{B}$ are minimal. Therefore, three different

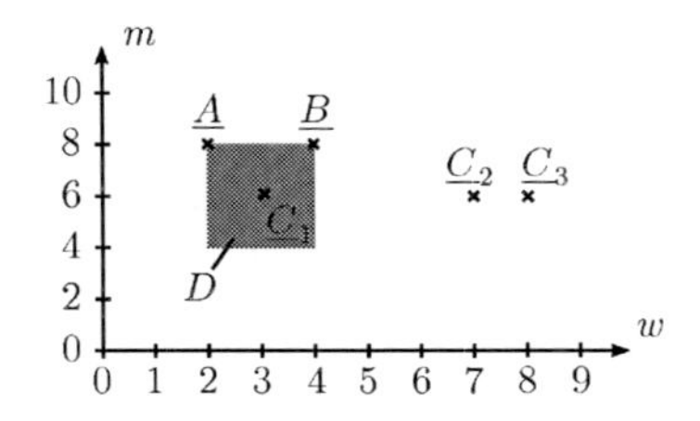

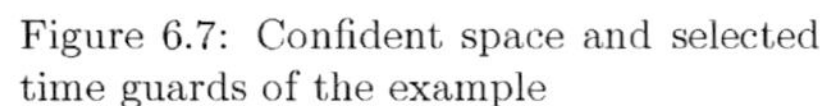

Figure 6.7: Confident space and selected time guards of the example

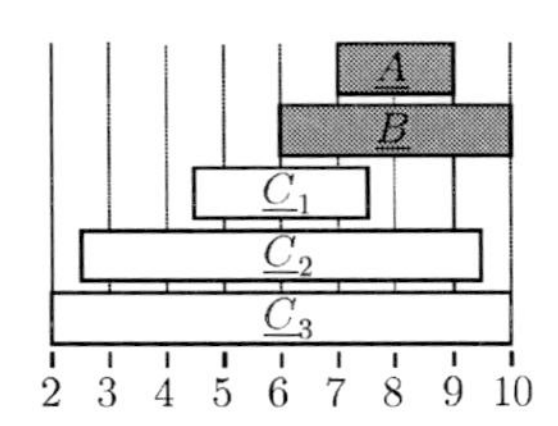

Figure 6.8: Timing intervals of the example

Table 6.2: Resulting rates and costs of the example

$\underline{T}_C$	$\underline{C}_1$		$\underline{C}_2$		$\underline{C}_3$	
$\underline{T}_R$	A	B	A	B	A	B
$MD(\underline{T}_R)$	0.83	0.50	0.71	0.50	0.75	0.50
$FA(\underline{T}_R)$	0.75	0.63	0	0.13	0	0
$h(\underline{T}_R, \underline{T}_C)$	**0.79**	0.56	**0.36**	0.32	**0.38**	0.25

intervals $\underline{C}_1 = (6,3)$, $\underline{C}_2 = (6,7)$, and $\underline{C}_3 = (6,8)$ are considered in the following. Their location within the (m,w)-domain is illustrated in Figure 6.7 and their alignment with respect to $\underline{A}$ and $\underline{B}$ in Figure 6.8. Note that a time guard must not necessarily lie within D. Given the weighting factor $K = 0.5$, the resulting costs for a selection of $\underline{T}_C$ according to Equation 6.16 are summarized in Table 6.2. The worst case costs $\max_{\underline{T}_R \in D}(h(\underline{T}_R, \underline{T}_C))$ for each $\underline{T}_C$ are marked with bold letters. One can see that the lowest worst case costs are obtained with $\underline{T}_C = \underline{C}_2$ and $\underline{T}_R = \underline{A}$. As a result, the interval $\underline{T}_C^* = \underline{C}_2$ is chosen as solution since it minimizes the worst case costs for the given selection of time guards.

6.2.4 Discussion

In this section, a selection of identification approaches for DES were presented that satisfy the requirements for *automatic* and *data-based* modeling. The proposed approaches generate models in a formal way using a finite set of system observations. The two most important advantages of data-based modeling compared to knowledge-based modeling are the *cost-efficiency*, due to the avoidance of manual interventions, and that the *real system behavior* is modeled instead of the *idealized* one [Roth et al., 2010]. During operation of a large and complex DES, unpredictable behavior may likely occur that is not specified but may nevertheless belong to the fault-free behavior of the system. This can result from external disturbances that affect sensors and actuators like dirt, moisture, and radiation, or from varying timings of controller calculations and communication delays. Typically, manual modeling does not consider these effects which may lead to a large amount of false detection during fault diagnosis. Model identification can deal with this issue since the data used for modeling reflects the accepted system behavior. However, the accuracy of an identified model heavily depends on the observation data. It has to be ensured that sufficient data is available containing the relevant system phenomena that are to be modeled.

An important requirement defined for the identification of DES is that the applied data are *passively* obtained observations of the *fault-free* system behavior. Active identification, as proposed in [Booth, 1967] for logical models and in [Grinchtein et al., 2005] for timed models, is not allowed. In the works of [Cabasino et al., 2007], [Verwer et al., 2006], and [Supavatanakul et al., 2006] negative observations of the system behavior or fault sequences are used. Since these observations do not belong to the fault-free system behavior, they can hardly be obtained by observing industrial systems. Some approaches need additional a priori information about the model or the system that is to be identified. This is often called *grey-box* identification. In [Cabasino et al., 2007], for instance, the maximum number of model states has to be a priori known, in [Basile et al., 2011] the exact timings of the events, and in [Lefebvre and Leclercq, 2011] and [Das and Holloway, 2000] the specific probability distributions. Since this knowledge is not available in general, the approaches are less suitable for the identification of industrial closed-loop DES.

Models that are used for fault diagnosis purposes have to satisfy requirements concerning precision and completeness. It is stated in [Klein, 2005] that most of the machine learning approaches typically cannot ensure this. Usually, it is only guaranteed that the observed behavior is reproduced by the identified model as in [Veelenturf, 1978] and [Meda-Campaña and López-Mellado, 2005]. If the entire system behavior is observed, the models can be completely identified but they may contain a significant degree of exceeding behavior. Identification approaches that provided guarantees for both, precision and completeness, are given by [Klein, 2005] and [Roth, 2010] for logical models and [Das and Holloway, 2000] for time behavior. However, the works consider only one single behavior aspect, either logical or time and do not provide a general timed model that can be applied for timed fault diagnosis.

The identification of real-world systems with *large size* poses additional challenges. In [Cabasino et al., 2007] and [Basile et al., 2011] an optimization problem has to be solved for identification whose complexity depends mainly on the size of the considered system. Such approaches are less suitable for large systems since computation demands become unacceptably high or a model may even not be found [Estrada-Vargas et al., 2010]. Other approaches, as [Supavatanakul et al., 2006], have only been validated for small academic examples and have never been evaluated with real-world systems.

6.3 Automatic Modeling of Concurrent DES

6.3.1 Preliminaries

Concurrent DES have been studied in literature in many fields ranging from control [Willner and Heymann, 1991] and model verification [Rohloff and Lafortune, 2002] to model-based fault diagnosis [Wang et al., 2007]. A commonly used approach to model concurrent system is a *distributed* model framework. The idea is to *partition* the model into a number of partial models such that each of them represents a concurrent operating subsystem of the DES, respectively. The concurrent behavior of the DES is then given by the combined behavior of all partial models. In [Philippot et al., 2007], this modeling concept is applied to DES in the field of fault diagnosis. The behavior of each subsystem is individually modeled by an automaton first and then the automata are combined to a global system model. A subsystem is represented by an actuator and its attached sensors. The behavior of these hardware components is assumed to be *causally related* since the operation of the actuator depends on signals provided by the sensors and vice versa. The partitioning of the model for all actuators and sensors is made by an expert using given knowledge about the structure and behavior of the controlled systems. Since the required information for manual partitioning of DES is typically not available, it has to be performed data-based using appropriate methods.

The following discussed methods for the automatic and data-based modeling of concurrent systems are divided into two categories. The first category contains approaches that extract *probabilistic information* from observed event sequences to discover concurrency. The approaches of the second category apply *model optimization* to find models that represent the concurrent behavior of systems in an optimal way. A selection of approaches of both categories is discussed in the following.

6.3.2 Probabilistic Data Mining

Workflow mining, also known as process mining, is a research topic in the field of business processes and enterprise information systems. A workflow is a sequence of tasks that are executed according to a defined process. The execution of a task is considered as an event. The idea of workflow mining is to recover a model of a given process in order to maintain or modify it. A possible application is to reveal discrepancies between the actual workflow processes and the processes specified by the management. A mining approach that focuses on the modeling of concurrent workflows is proposed in [Cook et al., 2004]. Specific measures such as *entropy*, *event type counts*, and *periodicity* are introduced in order to discover information about concurrent behavior from the observed event sequences. The measures rely on a *conditional probability* that is calculated using a frequency analysis of the event occurrences. This probability is introduced in [Cook et al., 2004] as:

$$CondProb(S) = \frac{Occur\,(S)}{Occur\,(Prefix\,(S))} \tag{6.19}$$

S is a sequence of events, $Occur\,(S)$ denotes the number of occurrences of S, and $Prefix\,(S)$ represents S with the last event removed. The conditional probability represents the frequency that the last element of S occurs under the condition that the first events of S are observed. For instance, assume that $Occur\,(ABC) = 75$ and $Occur\,(AB) = 100$, then $CondProb(ABC) = 0.75$. The result means, that event C follows the event sequence AB with probability 0.75 and with the remaining probability of 0.25 the sequence is followed by any other event. Based on the conditional probabilities, the *entropy* of a sequence S can be derived. The $Entropy(S)$ is an indicator for the distribution of probabilities among all following events of S. If a sequence S is always followed by an event A with $CondProb(SA) = 1.0$ and never by any other event e with $CondProb(Se) = 0.0\ \forall e$, then the result is $Entropy(S) = 0.0$. In this case, the behavior after S is deterministic. If the entropy is close to 1.0, several events follow S with equal probability. The metric can be used to determine synchronization points where two or more concurrent branches of event sequences start or end. If a synchronization point is found, the *event type counts* metric is used to distinguish between points that are followed by concurrent branches and points with sequential selections since both lead to similar entropy metrics. The *periodicity* metric is another measure that determines potential synchronization points in the model by searching for event sequences that are regularly recurring. Finally, all metrics are combined together to discover concurrency in a considered process.

In the work of [Estrada-Vargas, 2013], the concurrency of a DES is analyzed in order to identify a labeled Petri net that explicitly models causal and concurrency behavior. The identification algorithm consists of two basic steps. In the first step, the algorithm generates so-called *Petri net fragments* that model a part of the entire system behavior, respectively. The aim of the second step is to combine these sub-models to one monolithic Petri net such that the *causal* and *concurrency relationships* of the whole system are represented. In order to discover these relationships, the observed event sequences are translated into transition sequences of the Petri net fragments. These sequences can then be analyzed to obtain information about the relationships between transitions. A *causal relationship* is found if two transitions are observed consecutively such that one transition always precedes the other one. An example sequence for two transition t_a and t_b with a causal relationship is $\ldots, t_a, t_b, \ldots, t_a, t_b, \ldots$, in which the dots represent an arbitrary number of transitions that are different from t_a and t_b. A *concurrent relationship* is found if two transitions t_a and t_b are consecutively observed in both orders t_a, t_b and t_b, t_a while they have no causal relationship. This holds for instance in the following example sequence $\ldots, t_a, t_b, \ldots, t_b, \ldots, t_a, \ldots$. The author remarks that this condition for concurrency is very restrictive since it requires that all possible concurrent behavior of the system is observed. Due to that fact, several additional less constraining rules are proposed to determine concurrent relationship between the transitions of the Petri net fragments.

Another work, which used statistical information from observed event sequences to generate concurrent process models, is proposed in [Maruster et al., 2003]. The idea is to determine the relationships of events by a classification according to specific metrics. The three considered classes of relationships in this work are the *causal* c, the *concurrent* p, and the *exclusive relationship* e. The causal and concurrent relationships between two events are defined in the same way as in the preceding work of [Estrada-Vargas, 2013]. The exclusive relationship classifies another class of events that are neither causal related nor concurrent. This holds for two events that constitute a decision behavior within the system, for instance. Since one of the two events has to be chosen, they can never be observed consecutively or concurrently during a workflow. Five different metrics are calculated for each event pair based on the observed data. The metrics are called *causality metric* CM, *local metric* LM, *global metric* GM, *XY-metric* XY, and *YX-metric* YX. The results of the metrics are evaluated by a set of *rules* in order to classify the event pairs. The following rules are an excerpt from an exemplary rule base that is given in [Maruster et al., 2003]:

Rule1: IF $LM >= 0.949$ AND $XY >= 0.081$ THEN class c
Rule2: IF $LM >= 0.856$ AND $YX = 0$ AND $GM >= 0.224$ THEN class c
Rule3: IF $LM >= 0.844$ AND $CM >= 0.214$ AND $CM <= 0.438$ THEN class c

The first rule, for instance, classifies two events as causally related c if the values of the local metric LM and the XY metric satisfy certain bounds. A rule base typically consists of several dozens of rules. By evaluating the entire set of rules for each event pair with the corresponding metrics, all possible relationships are determined. They can then be used to identify the Petri net model. The key issue in the work of [Maruster et al., 2003] is the design of the rule base. It has to be determined in a way such that event relationships are classified correctly, even in the presence of noisy and incomplete data. As a solution, the authors propose to learn the rule base according to given examples of classified event pairs from known systems. The metrics of these examples are determined based on experimental data and the corresponding classification is done by an expert. In order to generate a representative and robust rule base, multiple training systems are considered and noise is artificially added to the training data. It is assumed that if the rules are learned in that way, many different aspects of system behavior are covered and the rules can be applied universally to discover concurrency in any arbitrary system. Further approaches for workflow mining and challenges in this field are presented in [van der Aalst and Weijters, 2004]. The survey highlights future research topics and actual problems such as how to make use time information or how to deal with noise and incomplete observation sets.

6.3.3 Model Optimization

The works discussed in the following formulate the automatic modeling of concurrent DES in terms of an *optimization problem*. In contrast to probabilistic data mining approaches, in which event sequences are analyzed to discover concurrency, model optimization focuses directly on the generation of models that can represent concurrent system behavior. The model optimization task is basically a search problem in this context. The aim is to find a model that represents the concurrent behavior of a DES in the optimal way with respect to certain quantitative criteria. In the work of [Medeiros et al., 2007], a *genetic algorithm* is proposed to solve this search problem. Genetic algorithms are heuristic optimization

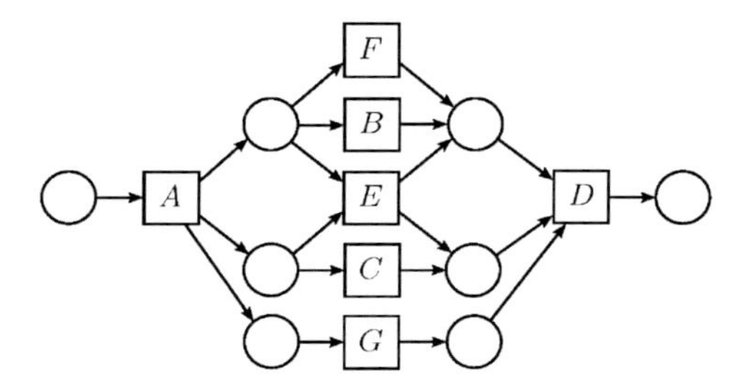

Figure 6.9: Example Petri net according to [Medeiros et al., 2007]

Table 6.3: Causal matrix of the example Petri net according to [Medeiros et al., 2007]

$ACTIVITY$	$I(ACTIVITY)$	$O(ACTIVITY)$
A	$\{\}$	$\{\{F,B,E\},\{E,C\},\{G\}\}$
B	$\{\{A\}\}$	$\{\{D\}\}$
C	$\{\{A\}\}$	$\{\{D\}\}$
D	$\{\{F,B,E\},\{E,C\},\{G\}\}$	$\{\}$
E	$\{\{A\}\}$	$\{\{D\}\}$
F	$\{\{A\}\}$	$\{\{D\}\}$
G	$\{\{A\}\}$	$\{\{D\}\}$

strategies using concepts inspired by natural selection such as elitism, crossover, and mutation. The algorithm uses these mechanisms to generate new solutions with increasing fitness values towards the optimal solution. The fitness value represents the quality of a found solution while the optimal solution has typically the highest value. The most important concerns using genetic algorithms for solving optimization problems are the *internal representation* of the search space and the *fitness measure*. In the approach of [Medeiros et al., 2007], which aims to find a Petri net model, an internal representation called *causal matrix* CM is applied. This matrix contains the dependencies between all places and transitions and can be transformed into an equivalent net. In Figure 6.9 and Table 6.3, an example Petri net from [Medeiros et al., 2007] and the corresponding causal matrix is depicted. The events of the work flow are called activities. Each $ACTIVITY$ has an input set $I(ACTIVITY)$ of activities that have to occur in order to enable $ACTIVITY$ and an output set $O(ACTIVITY)$ of activities that are enabled upon the occurrence of $ACTIVITY$. To observe activity B in the example, one can see in Figure 6.9 that it is required to observe activity A before. Activating B, on the other hand, is a precondition for the activation of D. This relationship refers to the second row in Table 6.3. The fitness measure is formulated with respect to requirements for model *completeness* and *precision*. In [Medeiros et al., 2007], completeness is defined with respect the observed behavior in the recorded data. A model is assumed to be complete if it is able to reproduce the complete observed behavior. The same holds for precision that is defined in [Medeiros et al., 2007] as the exceeding behavior of the model with respect to the observed one.[15] Two partial fitness functions are introduced, one to ensure completeness and one to ensure precision of a generated solution, respectively. The function that addresses the completeness of a found solution for a causal matrix CM based on observed sequences L is given in [Medeiros et al., 2007] as:

$$PF_{complete}(L, CM) = \frac{allParsedActivities(L, CM) - punishment(L, CM)}{numActivitiesLog(L)} \tag{6.20}$$

$allParsedActivities(L, CM)$ denotes the number of observed events that can be reproduced by a solution, $numActivitiesLog(L)$ represents the number of events in L, and $punishment(L, CM)$ is a penalty for a solution that does not correctly represent the observed behavior. In the best case, the solution is able to reproduce the entire observed behavior and the value of the partial fitness function is $PF_{complete}(L, CM) = 1$. The second partial fitness function addresses the precision of a found solution with respect to the observed behavior. It is assumed that models that absolutely perform less behavior are preferable since they are more adapted to the specific behavior of the considered DES. The amount of behavior a model can perform is defined as the number of events that are activated in each state, i.e. the number of transitions in the PN that may fire. The function to determine this partial fitness of a causal matrix CM among other possible causal matrices $CM[\,]$ is given in [Medeiros et al., 2007] as:

$$PF_{precise}(L, CM, CM[\,]) = \frac{allEnabledActivities(L, CM)}{max(allEnabledActivities(L, CM[\,]))} \tag{6.21}$$

$allEnabledActivities(L, CM)$ is the number of events that may be executed based on CM during the execution of the observed sequences L and $max(allEnabledActivities(L, CM[\,])$ is the maximum value of activated events for all given causal matrices $CM[\,]$. Models that allow much extra behavior are usually defined in a too general way. Their partial fitness $PF_{precise}(L, CM, CM[\,])$ has a value close to 1 while specialized model have partial fitness close to 0. To balance the demands on completeness and precision, the final fitness measure combines both partial fitness functions by a selectable parameter $\psi \in [0, 1]$. The equation for the combined fitness measure is formulated in [Medeiros et al., 2007] as:

$$F(L, CM, CM[\,]) = PF_{complete}(L, CM) - \psi \cdot PF_{precise}(L, CM, CM[\,]). \tag{6.22}$$

The causal matrix with the highest fitness value $F(L, CM, CM[\,])$ is the solution of the optimization problem. The Petri net that is equivalent to this causal matrix is the optimal model with respect to the here defined completeness and precision requirements.

In [Roth, 2010], an optimization strategy for *partitioning* of concurrent closed-loop DES models is proposed. The aim is to automatically generate a set of partial models $sys_1, sys_2, \ldots, sys_n$ such that each partial model sys_i represents a concurrent operating

[15] The definition of precision and completeness given in [Medeiros et al., 2007] differs from the definitions introduced in Chapter 2.

part of the DES, respectively. A partial model sys_i is defined according to a subset of controller I/Os, denoted by $y(sys_i) \subset I/O$, while all partial models are independent of each other $y(sys_i) \cap y(sys_j) = \emptyset$, $i \neq j$ if no additional a priori information during optimization is used. It is assumed in this work that the number of required partial models N_{sys} is available and that a sufficient number of observations of the fault-free system behavior is given. As in the preceding work, the optimization constitutes a global search problem. To find a solution, all partial models have to be generated simultaneously. A *simulated annealing* algorithm is proposed to solve this global problem. Simulated annealing is a heuristic optimization method inspired by the annealing process in metallurgy. A optimization run starts with an initial set of partial models and a given high temperature. The temperature of the system is continuously decreased while new generated solutions are accepted with a probability that depends on their fitness and on the current temperature of the system. If a new solution is obtained, the partial models are collectively evaluated with a global fitness criterion. The applied criterion focuses exclusively on the completeness of the resulting partial models. A criterion that considered the precision of a resulting model is not formulated explicitly. However, in order to minimize the exceeding language, the number of resulting partial models N_{sys} should be kept as small as possible.

6.3.4 Discussion

Automatic modeling of concurrent DES can basically be interpreted in different ways. It can either be considered as an issue that complicates model identification as shown in [Estrada-Vargas, 2013] and [Medeiros et al., 2007] for instance, or as a *partitioning task* that precedes the actual model identification as in [Roth, 2010]. However, both considerations focus on discovering system concurrency in order to adapt the model in an appropriate way. The basic intention of workflow mining approaches is to recover process models based on observed data that can be interpreted and analyzed by experts. Usually, Petri nets are used since they allow a compact and clear representation of sequential and concurrent behavior. If models are generated for FDI purposes, a clear representation is not required. Rather the most important concerns are the generation of *precise* and *complete* models while the computation demands are bounded even with large systems. Probabilistic data mining approaches can usually not give guarantees for precision and completeness. These approaches use specific metrics as in [Cook et al., 2004] and [Maruster et al., 2003], or heuristic rules as in [Estrada-Vargas, 2013] to discover event dependencies. They rely on *local* information that is given by the consecutive occurrence of events. Since it is very likely that not all event dependencies can be captured in this way, the generation of complete and precise models is hardly possible. Furthermore, probabilistic approaches are highly sensitive against noisy and incomplete observed data. On the contrary, their advantages are low computational efforts and interpretable results. This makes probabilistic data mining very appropriate to determine basic relationships between events and to generate models for applications that make no demands on precision and completeness.

Model optimization approaches ensure the quality of a model by developing appropriate fitness criteria as in [Medeiros et al., 2007] or by additional optimization constraints

[Roth, 2010]. The criteria are explicitly formulated with respect to the precision and completeness properties. The optimization approaches perform a *global* search based on the observed data instead of generating only local information. Compared to probabilistic approaches, they are robust against noise and incomplete observation data [Medeiros et al., 2007]. One of the most challenging issues in the context of model optimization is the *computational complexity*. In the work of [Medeiros et al., 2007], the computation of the optimization problem becomes difficult for a system that contains 22 different events. This shows that the approach is limited to small systems that have only a few hardware components. The optimization approach of [Roth, 2010] is developed for large closed-loop DES and has been successfully applied to real-world industrial systems. However, the interaction of subsystems via shared I/Os cannot be handled without additional expert knowledge since the computational effort of the optimization would become impracticably large. Instead, the approach always generates partial models that operate independent of each other. The resulting model is complete but in general too permissive with respect to the original behavior. In order to minimize the exceeding behavior, the synchronization needs to be explicitly considered during the optimization process or additional concepts for behavior restriction, as proposed by [Roth, 2010], must be used.

6.4 Model-based Fault Diagnosis of DES

6.4.1 Preliminaries

Many different concepts for model-based FDI of DES have been proposed by the scientific community. The capabilities of these concepts are closely related to the type of system model that is applied. According to this, the FDI concepts introduced in the following are divided into two main categories. The first category contains concepts that use *logical behavior models* for the diagnosis of logical faults The second category contains concepts using models that additionally model the *time behavior* of a DES for timed FDI. Models of both categories may represent the *fault-free* or *faulty* behavior of the considered system, as discussed in Chapter 2. Another characterization of model-based fault diagnosis approaches is to distinguish between *centralized*, *decentralized* and *distributed* diagnosis. While centralized approaches perform global fault diagnosis of the entire system using a monolithic model, decentralized and distributed perform local fault diagnosis, see [Zaytoon and Lafortune, 2013]. The difference between decentralized and distributed approaches is that decentralized diagnosis relies on a global model, while distributed diagnosis uses a set of sub-models with communication between the local diagnosis units. The units may exchange local diagnosis information to improve the overall diagnosis result. Comprising surveys about fault diagnosis concepts for DES using automata and Petri net models are presented in [Zaytoon and Lafortune, 2013] and [Fanti and Seatzu, 2008]. In the following, selected approaches are explained and discussed with respect to the challenges introduced in Chapter 2 for industrial closed-loop DES.

6.4.2 Fault Diagnosis using Logical Models

A well-known approach for model-based fault diagnosis of DES is the diagnoser approach, published in [Sampath et al., 1995]. The scientific community pays much attention to this work since it provides a well-proven and systematic formalism for modeling and fault diagnosis. The main contribution of this work is the design and use of an observer automaton, called *diagnoser*. This automaton can be applied for online fault diagnosis and to verify the *diagnosability* property. In order to develop a diagnoser, the fault-free and faulty behavior of the system needs to be modeled. In [Sampath et al., 1996], the authors explain by means of an example how this can be done using detailed knowledge of the considered system. The procedure consists of building automata models for the plant components and the controller, respectively. Then, these automata are composed to a global model G by using standard composition operations for automata. The resulting automaton G contains an event set $\Sigma = \Sigma_o \cup \Sigma_{uo}$ that is used to represent the fault-free and faulty behavior of the controlled DES. The set contains the observable Σ_o and unobservable events Σ_{uo}, while fault related events are typically modeled as unobservable $\Sigma_f \subseteq \Sigma_{uo}$. Based on G, the diagnoser automaton can then be derived. To perform online fault diagnosis, the diagnoser observes the behavior of the system by means of the observable events. The idea is to infer the occurrence of unobservable fault events by evaluation of the observable ones. In order to proof that it is possible to diagnose all faults within a finite time period the diagnosability property is introduced. Diagnosability is defined in [Cassandras and Lafortune, 2008] as:

Definition 38 (Diagnosability). "Unobservable event e_d is *not diagnosable* in live language $L(G)$ if there exist two strings s_N and s_Y in $L(G)$ that satisfy the following conditions: (i) s_Y contains e_d and s_N does not; (ii) s_Y is of arbitrarily long length after e_d; and (iii) $P(s_N) = P(s_Y)$. When no such pair of strings exist, e_d is said to be *diagnosable*".

In other words, if there exist two arbitrary long sequences s_N and s_Y in the diagnoser, one without and one with the fault event e_d and both sequences s_N and s_Y represent the same sequence of observable events, then e_d is not diagnosable. The projection operation $P(s)$ is used to erase unobservable events from an event sequence s, which consists of observable and unobservable ones. Non-diagnosability typically refers to faults that do not cause observable fault symptoms. In this case, the system behavior after the occurrence of a fault corresponds to the system behavior in the fault-free case. A system is said to be diagnosable if all modeled fault events are diagnosable according to Definition 38. The diagnosability property is not restricted to a specific model definition, it is rather a general property of fault models that are used for model-based fault diagnosis purposes.

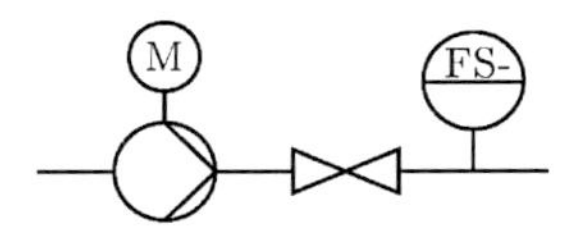

Figure 6.10: Pump-valve example system

To illustrate the concept diagnoser and diagnosability, the example system, illustrated in Figure 6.10, is considered. The figure shows the piping and instrumentation diagram of a simple pump-valve system with an attached sensor that indicates a flow through the connecting pipe. The system is controlled and has four observable events $\Sigma_o = \{start, stop, open, close\}$ representing the actuating commands. The events *start* and *stop* control the pump while *open* and *close* are used to actuate the valve. Two unobservable events $\Sigma_{uo} = \Sigma_f = \{stuck\_close, stuck\_open\}$ represent two different fault situations of the valve. The fault event *stuck_close* indicates that the valve is closed and can no longer be opened while *stuck_open* represents the opposite case that the valve is open and can no longer be closed. During normal, fault-free operation of the system, a fluid flows through the pipe if the valve is opened and the pump is switched on. This is indicated by the sensor measurement FL. If either the valve is closed or the pump is switched off, then there is no flow NF. In the initial state of the system, the pump is switched off and the valve is closed. The controlled system can switch between the operation modes (flow and no flow) by stopping the pump and closing the valve or opening the valve and starting the pump, respectively.

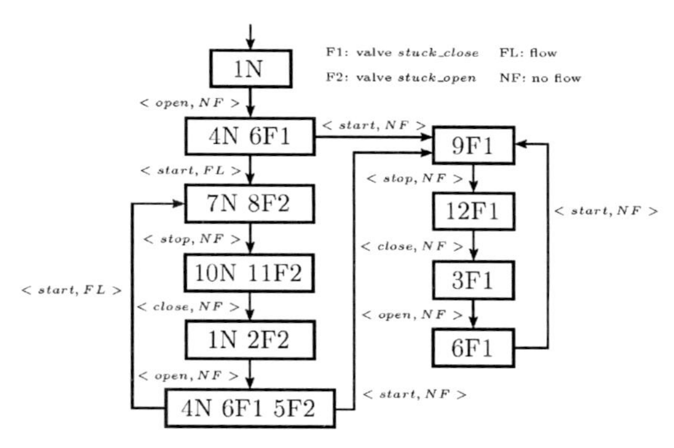

Figure 6.11: Diagnoser for the pump-valve example system

In Figure 6.11, the diagnoser automaton for the pump-valve example is depicted. Each state of the diagnoser is labeled with a set of possible model states given by the numbers and a corresponding fault label. The state numbers i refer to the states x_i of the observed combined automaton that can be reached so far based on the observed events. The fault labels of each state contain the diagnosis information. States that correspond to the normal behavior are labeled with N while states that correspond to a faulty behavior of type j are labeled with Fj. In this example, the fault label $F1$ indicates that the valve is *stuck_close* and the fault label $F2$ indicates that the valve is *stuck_open*. If only fault labels occur in a diagnoser state, the system is definitely in a fault-state. If a diagnoser state contains the N label and one or more fault labels, it cannot be decided if the system is fault-free or faulty. In this case, both conditions are possible based on the observations that are made so far. For instance, in the first state of the diagnoser there is the label $1N$, which means that the underlying model state x_1 is reached and the system runs normal. The second state of the diagnoser contains two models states x_4 and x_6. The model state x_4 corresponds to the fault-free operation and state x_6 to the valve stuck close fault. In this case, it cannot be decided if the system is operating fault-free or fault $F1$ has happened. The transitions of the diagnoser are triggered by observable events and the corresponding sensor information. The transition from the initial to the second state of the diagnoser, for instance, is performed based on the observed event *open* and the sensor information NF. A sequence of events that correspond to the fault-free behavior can be given as

$$\sigma_n =< open, NF >, < start, FL >, < stop, NF >, < close, NF >, < open, NF >, \ldots.$$

Starting in the initial state, one can see that the state trajectory of the diagnoser resulting from σ_n consists of states that contain the N label. A sequence that corresponds to a *stuck_close* fault $F1$ of the valve is given as

$$\begin{aligned}\sigma_{F1} = &< open, NF >, < start, FL >, < stop, NF >, < close, NF >, \underline{< stuck\_close >}, \\ &< open, NF >, < start, NF >, \ldots.\end{aligned}$$

The underlined fault event is unobservable and cannot be recognized by the diagnoser. However, by observing the $< start, NF >$-event after the fault event has occurred, the diagnoser enters the $9F1$ state. In this state, the $F1$ fault is unambiguously diagnosed since the state contains no N label and no other fault is possible. A possible sequence based on a *stuck_open* fault $F2$ is

$$\begin{aligned}\sigma_{F2} = &< open, NF >, \underline{< stuck\_open >}, < start, FL >, < stop, NF >, < close, NF >, \\ &< open, NF >, \ldots.\end{aligned}$$

After the fault event occurs, the diagnoser enters a cycle of states that is equal to the cycle entered upon the occurrence of the fault-free event sequence σ_n. One can see that if the fault event is deleted in σ_{F2}, then the cycles are identical $P(\sigma_{F2}) = P(\sigma_n)$. As a result, the fault $F2$ cannot be diagnosed since it cannot be decided based on the observations whether the system is in the normal state N or in the fault state $F2$. The fault $F2$ is not diagnosable.

The diagnoser, introduced in [Sampath et al., 1995], represents a *centralized* fault diagnosis approach. It is assumed that the applied diagnoser has access to all observable events of the DES. However, in some cases the DES is too large to be diagnosed with one single diagnoser or the events are only *decentralized* available among several sites of the system. Examples for systems like this are local work stations in a communication network or a decentralized controlled manufacturing process. For these kind of systems, a decentralized diagnoser approach is presented in [Wang et al., 2007]. A system G, represented by a finite state automaton, is diagnosed by multiple, local diagnosers. Figure 6.12 shows the structure of this decentralized diagnosis system. Each local diagnoser D_i observes a subset of all observable event $\Sigma_{o,i} \subset \Sigma_o$. The projection operation P_i assigns

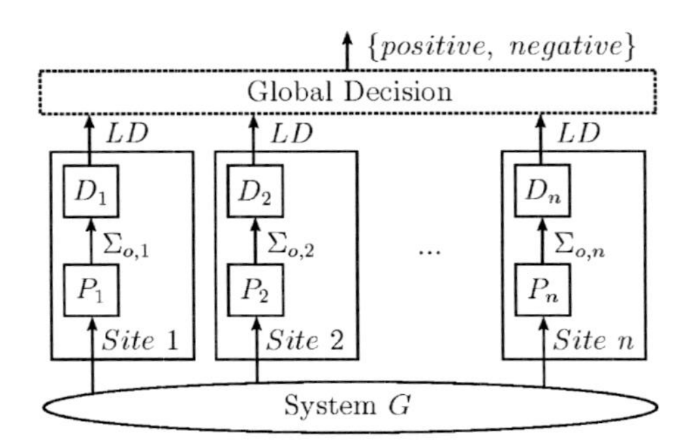

Figure 6.12: Decentralized diagnoser concept [Wang et al., 2007]

the corresponding events from Σ_o to $\Sigma_{o,i}$. A local diagnoser calculates a local decision LD and forwards it to a global coordinator in order to fuse all local results to a global diagnosis decision. The result of the global decision is either *true* if a fault is detected or *false* if not. There is no communication among the local diagnosers and the coordinator has no knowledge about the system model. The coordinator in this approach consists of simple Boolean algebra. It can be implemented straightforward and requires only little memory and processing capabilities. In order to improve the diagnosis performance, the authors of [Debouk et al., 2000] propose several advanced protocols for a decentralized diagnoser. Each protocol defines the diagnostic information generated at the local diagnoser, the communication rules between the local diagnosers, and the coordinator and the decision rule of the coordinator. The protocols differ in diagnosis performance, system restrictions, communication efforts, computation requirements, and implementation complexity. It is shown that using these protocols, a decentralized diagnoser can perform as good as the corresponding centralized under certain conditions.

In [Dotoli et al., 2009], a fault detection approach is proposed that is based on a centralized Petri net model of the fault-free and faulty system behavior. It is assumed that the Petri net model of the considered DES and the initial marking is given. The behavior of the system is represented by observable and unobservable transitions in the model while faults correspond to a subset of the unobservable ones. The idea of the online fault detection algorithm is to check whether an observed sequence of events is consistent with the modeled normal behavior of the system or with any of the modeled faulty behavior. This check is formulated in terms of *ILP problems*. For each fault, sequences of transitions are determined that maximize the occurrence of the fault event under the condition that the sequence of observed events is executed. A solution sequence corresponds to the observed behavior and contains the considered fault event at least one time. If such a sequence of transitions can be found, the considered fault is added to the list of possible faults. If no such sequences can be found for any fault, the behavior of the system is considered as normal. The authors remark that no polynomial algorithm for solving ILP problems exists. Hence, the computational efforts of the proposed approach can be considerably high. Another model-based fault diagnosis approach that is based on a Petri net model is presented in [Genc and Lafortune, 2003]. The *distributed* diagnosis approach relies on the diagnoser concept proposed in [Sampath et al., 1995]. Two Petri net diagnosers are used to detect and isolate faults while they are communicating with each other. The diagnosers share information about change of markings and update their own markings and diagnosis information correspondingly.

The theory of *diagnosis from first principles*, presented in [Reiter, 1987], is a general study on fault diagnosis of systems using a fault-free system description. The idea is that only a description of the system SD and of the system components $COMPONENTS$ is available. This description specifies how the system normally behaves when all system components are operating fault-free. If an observation OBS of the system is consistent with the description of the system and the functioning of the components, then the system is assumed to be fault-free. This proposition is formally describe in [Reiter, 1987] as:

$$SD \cup OBS \cup \{\neg AB(c) \mid c \in COMPONENTS\} \tag{6.23}$$

with $\neg AB(c)$ denoting the normal, fault-free functioning of component c. In case of a fault, the diagnosis problem is to determine the abnormal functioning components such that the discrepancy between the observed and described system behavior can be explained. It is assumed that a fault leads to an observed system behavior OBS such that Equation 6.23 no longer holds. The isolation of the faulty components is based on the *Principle of Parsimony*, which says that "a diagnosis is a conjecture that some minimal set of components are faulty" [Reiter, 1987]. The consistency in case of a fault can then be defined, according to [Reiter, 1987], as:

$$SD \cup OBS \cup \{AB(c) \mid c \in \Delta\} \cup \{\neg AB(c) \mid c \in COMPONENTS - \Delta\} \tag{6.24}$$

with $AB(c)$ denoting the abnormal, faulty functioning of a component and Δ is the minimal set of faulty components. The faulty components of Δ are assumed to be the cause for the observed abnormal behavior since it is in general more likely that few rather than many components are faulty at the same time. However, the result of fault diagnosis is not unique. It is in general possible to obtain more than one diagnosis result that is able to explain the observed abnormal behavior of the system.

In the work of [Roth, 2010], a distributed fault diagnosis approach is developed for large and concurrent closed-loop DES. The behavior of these controlled systems is captured by events produced by the controller I/Os. If one or more I/Os change its values from discrete time step k to $k+1$, the system raises an event. For online fault diagnosis, the behavior of all I/Os is monitored. As in the preceding work, the applied model represents the fault-free behavior of the system. To deal with concurrency, a distributed modeling structure, called *automata network*, is used. An automata network is a collection of n partial automata, each representing the behavior of a subsystem, respectively. Fault diagnosis is performed by comparing the observed with the modeled logical behavior. If the observed behavior is not consistent with the modeled one, a fault is detected and can be isolated.

The combined behavior of all partial automata models the behavior of the concurrent system. Since the partial models are typically independent of each other and time behavior

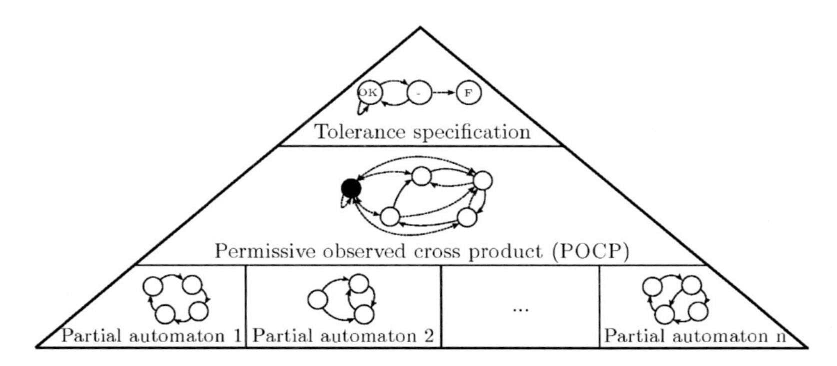

Figure 6.13: Restricted automata network for fault diagnosis [Roth, 2010]

is not considered, the modeled behavior is in general too permissive. This can lead to missed detections. In order to restrict the behavior of the automata network, the author introduces the *permissive observed cross product* (POCP) and the *tolerance specification*. This is depicted in Figure 6.13. The POCP is basically a parallel composition of the partial automata. After the composition a distinction is drawn between transitions that correspond to observed behavior and transitions that have not been observed but result from the composition operation. In the figure, the observed transitions are indicated by solid edges and unobserved ones by dotted ones. The unobserved transitions typically originate from the system concurrency. They most likely refer to fault-free system behavior but it cannot be guaranteed in general. In order to reproduce further unobserved behavior, an additional joker state, which connects all the other states by unobserved transitions, is added to the POCP. This state is the solid black one in the figure. The tolerance specification defines how many unobserved transitions may be executed before a fault F is declared. The idea is that a sequence of some unobserved transition does not necessarily represent a faulty behavior if the following observations are consistent with the modeled behavior again. In this case, no fault should be declared. However, if a significant number of unexpected observations are made in a sequence, the behavior of the system is most likely faulty. The specification is manually defined allows to make a trade-off between unknown behavior that is not modeled and undesired exceeding behavior.

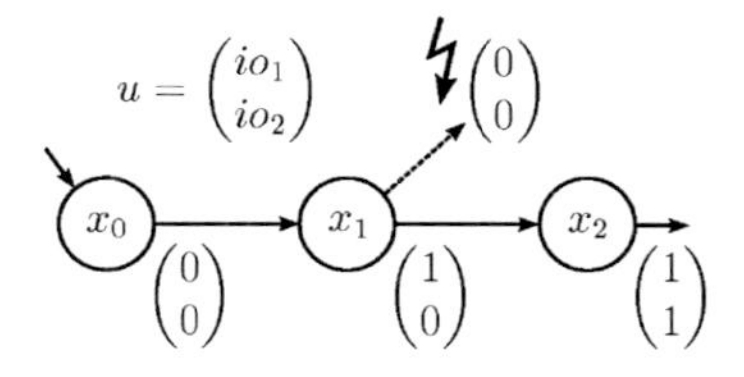

Figure 6.14: Example for a detected fault due to a logical inconsistency

For the isolation of detected faults, specific DES residuals are introduced in [Roth, 2010]. The residuals determine logical fault symptoms based on the detected inconsistency between the observed and modeled logical system behavior that leads to fault detection. The considered fault symptoms are *unexpected behavior* and *missed behavior*. Unexpected behavior is behavior generated by a DES that is not expected with respect to the modeled behavior. Missed behavior represents modeled behavior that is expected but not observed. By means of the residual, a set of sensors and actuators can be isolated that are related to the detected fault.

An example for the unexpected and missed behavior residuals based on a monolithic model is presented in the following. Consider the automaton given in Figure 6.14. Given the current model state $\tilde{x} = x_1$, a new observation

$$u(t) = \begin{pmatrix} 0 \\ 0 \end{pmatrix}$$

is made that is inconsistent with the modeled behavior since no following state of x_1 exists that can reproduce the observed behavior. Hence, a fault is detected. The unexpected behavior residual $Res1$ and the missed behavior residual $Res3$ are determined as

$$Res1(\tilde{x}, u(t)) = \{io_1\downarrow\} \quad \text{and} \quad Res3(\tilde{x}, u(t)) = \{io_2\uparrow\}$$

with $io_1 \downarrow$ denoting the falling edge of io_1 and $io_2 \uparrow$ denoting the rising edge of io_2, respectively.[16] The results show that for the given observation the changing signal of io_1 from 1 to 0 occurred unexpectedly while the changing signal of io_2 is missed with respect to the modeled behavior. Consequently, the sensors and actuators connected to both I/Os are likely related to the detected fault. Additional residuals $Res2$ and $Res4$ determine more possible fault candidates in a similar way.

6.4.3 Fault Diagnosis using Timed Models

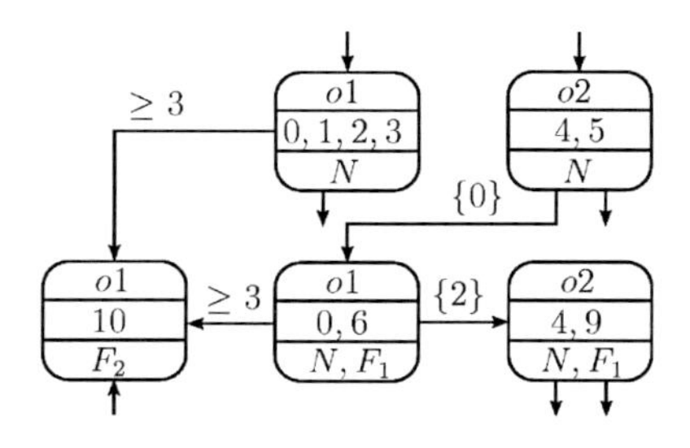

Figure 6.15: Section of the timed diagnoser example in [Hashtrudi Zad et al., 2005]

In [Hashtrudi Zad et al., 2005], a timed extension of the classical diagnoser approach after [Sampath et al., 1995] is presented. It is assumed that timing information of the

[16] See Section 5.4 for the definition of rising and falling edges of I/Os.

fault-free and faulty system behavior is given in the system model that can be used to construct the timed diagnoser. The advantage of adding time information is that time related faults can be detected and isolated such as deadlocks, for instance. The progress of time is modeled by time events. With each tick of a global clock an observable time event τ is raised. The logical behavior is modeled by the evolution of state outputs. Each state x is associated with an output $\lambda(x)$. The state estimation is performed by observing changing outputs $\lambda(x) \neq \lambda(x')$ from state x to a following state x'. All events of the DES are assumed to be unobservable with the exception of the time tick event τ. A section of a timed diagnoser is depicted in Figure 6.15. As in the original definition, each diagnoser state contains the estimated states of the underlying automaton model, denoted by numbers and the fault labels N and F_i. Additionally, each diagnoser state models an output o_j that is the common output of the estimated model states. The transitions between the diagnoser states are defined with respect to time constraints. The constraints are either given in terms of sets, whose elements represent the number of time events that must occur before the transition fires, or as inequalities, e.g. ≤ 3. An example of an observed state output trajectory with respect to the diagnoser given in Figure 6.15 is: $\ldots o_2 \xrightarrow{e} o_1 \xrightarrow{\tau} o_1 \xrightarrow{\tau} o_1 \xrightarrow{e} o_2 \ldots$ with $e \in \Sigma$ as an unobservable event that cases the DES to change its state. Assume that the diagnoser state $(o_2, (4,5), N)$ is active first, then the event e that instantaneously occurs leads the diagnoser to change its state to $(o_1, (0,6), (N, F_1))$. The state transition can be observed by the changing output. Two time ticks later, the event e leads the system to enter a state with a differing system output o_2. This, again, activates the diagnoser state $(o_2, (4,9), (N, F_1))$. As a remark, the output does not necessarily change on a state transition. For instance, on executing the transition from $(o_1, (0,6), (N, F_1))$ to $(o_1, 10, F_2)$, constrained by the time condition ≤ 3, the output o_1 remains the same. Fault diagnosis approaches, like this one, that rely on the evolution of states are often called *state-based diagnosis* in literature [Zaytoon and Sayed-Mouchaweh, 2012]. In [Philippot et al., 2007], timed fault diagnosis using timed diagnosers is combined with a *decentralized* model structure. Multiple local diagnosers perform local fault diagnosis for a defined part of the system. The global fault diagnosis decision is obtained by merging the local decision into a global one using the Boolean 'union' operator.

A timed automaton is applied in [Supavatanakul et al., 2006] for timed fault diagnosis of a DES. The automaton models the logical and timed behavior of the fault-free and faulty operation of the system. By reproducing the observed behavior with the modeled one, reachable states are identified that belong either to the fault-free or faulty behavior of the system. The fault diagnosis problem is formulated in [Supavatanakul et al., 2006] as:

$$\mathcal{F}(V(0 \ldots t_h), W(0 \ldots, t_h)) = \{f(t_h) \mid Poss(f \mid V(0 \ldots t_h), W(0 \ldots t_h)) = 1\} \quad (6.25)$$

$V(0 \ldots t_h)$ and $W(0 \ldots, t_h)$ are observed I/O sequences up to the time horizon t_h, respectively, and $f(t_h)$ is a possible fault f that occurred within t_h. In words, the aim of diagnosis is to determine a set of faults that may have happened based on the observed sequences. Since the automaton is non-deterministic, the set can contain several faults rather than only one. Then, the decision which fault has happened cannot be unambiguously made. In order to determine the set of possible faults, the *state observation problem* needs to be solved. The task consists of determining the states of the automaton that can be reached based the observed I/O sequences. This problem is in [Supavatanakul et al., 2006] formally given as:

$$\mathcal{L}_{dia}(t_k) = \{z_{dia}(t_k) \mid Poss(Z_{dia}(0 \dots t_k) \mid V(0 \dots t_k), W(0 \dots, t_k)) = 1\} \qquad (6.26)$$

with $t_k \leq t_h$ and z_{dia} is a fault interpreted state of the automaton that is reachable after t_k observed I/O sequences. The state estimation initially considers all states of the automaton as possible initial states. With ongoing observation t_k, the set of reachable states is updated such that the logical and time observations are consistent with the modeled behavior. By extracting the fault information from the estimated states, the possible faults are determined and added to $\mathcal{F}(t_k)$. In order to illustrate the state estimation and fault diagnosis procedure, an example is considered based on the timed automaton shown in Figure 6.6. It is assumed that x_0, x_1, x_2, x_3, x_4 correspond to the normal behavior, indicated by f_0, and x_0, x_1, x_2, x_5 correspond to fault f_1. In the initial step, the complete state-space is estimated since no knowledge about the initial state is available. Hence, the estimated faults are all possible ones $\mathcal{F}(0) = \{f_0, f_1\}$. After the first observation ($v = 1, w = 2$), the state estimation is given as $\mathcal{L}_{dia}(1) = \{x_0, x_2\}$. These are the only states that can be reached in the automaton model based on the observed I/O values. The faults that belong to these states are $\mathcal{F}(1) = \{f_0, f_1\}$. The second observation ($v = 1, w = 2$) leads to $\mathcal{L}_{dia}(2) = \{x_5\}$ and $\mathcal{F}(2) = \{f_1\}$. After this observation, the model state and the fault can be unambiguously determined.

In [Jiroveanu and Boel, 2006] a distributed approach for timed fault diagnosis of DES is proposed. It is assumed that the considered system is composed of partial systems that are interacting with each other. Each partial system is modeled by a TPN. The partial models are observed by so called *agents* that compute local fault diagnosis. The concept of agents is similar to the concept of local diagnosers for Petri nets that are presented in [Genc and Lafortune, 2003], for instance. The task of an agent is to estimate all sequences of transitions that can fire based on the observed events in order to conclude whether the partial system is operating normal, a fault has happened, or no unambiguous decision about the condition can be made. Faults are modeled in the system by unobservable events attached to transitions. An important contribution in this work is the interaction between two or more partial models represented as *synchronizing transitions* and *logical guards*. Synchronizing transition are commonly shared transitions among several partial TPNs. Logical guards are additional conditions that restrict the logical behavior of a partial model besides the partial events. Additionally, time guards at transitions are applied to constrain the timed model behavior.

In Figure 6.16, an example from [Jiroveanu and Boel, 2006] for two interacting submodels $Comb_1$ and $Comp_2$ is depicted. The interaction is given by the logical guard $G := (m(p_5) = 1)$ that is attached to transition t_2 in $Comb_1$. The guard G is true if place p_5 in $Comp_2$ is marked. The transition t_2 is fires if the event belonging to t_2 occurs while the logic guard G and time guard $[5, 10]$ are satisfied. In order to ensure that G is true, $Comp_2$ must perform a transition sequence based on a specific timing such that p_5 is

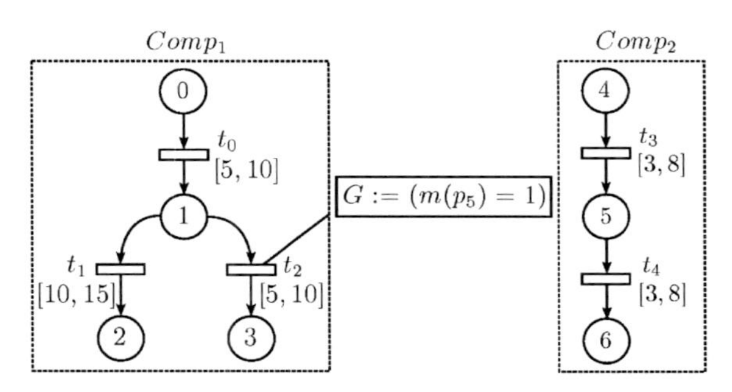

Figure 6.16: Example of interacting local Petri nets [Jiroveanu and Boel, 2006]

marked right in time. Synchronization transitions and logical guards pose restrictions on the behavior of all involved partial models. Since all local models must be aware of that local constraints, they are communicated among the agents at defined times during fault diagnosis. This information helps to refine the local estimations of transition sequences and ensures that the local behavior of a partial model is consistent with the behavior of the other local models. If a global consensus can be achieved, the local diagnosis decisions can then be combined to recover the global diagnosis result.

In [Pandalai and Holloway, 2000], an approach for fault detection is proposed that uses a *template model* to represent the fault-free behavior of a system. Template models define possible sequences of events and their corresponding timings. The state of a template model is given by a set of expectations $O(t)$. The set $O(t)$ is a collection of events with timing constraints that are expected to occur in future. Fault detection is performed by evaluating the expectations. If either no expectation is fulfilled or an unexpected event is observed, a fault is declared. With ongoing observation of the system, expectations are added to $O(t)$ according to the defined template model when events occur and removed from $O(t)$ when they become satisfied or violated. An expectation is given as (t, e, C, w) with t the time when e is observed, e is the triggering event, C is the set of consequences, and w is a label that is used for the monitoring procedure. A consequence is given as a pair (e', τ), e' is the consequence event and τ is the time constraint given in an interval notation.

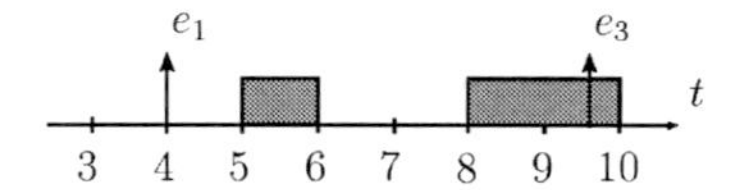

Figure 6.17: Example of a satisfied expectation

In the following, an expectation is given as $(4, e_1, \{(e_2, [1, 2]), (e_3, [4, 6])\})$. The expectation is triggered by the event e_1 that is observed at $t = 4$ and contains two consequences, namely $(e_2, [1, 2])$ and $(e_3, [4, 6])$. The expectations is satisfied if either e_2 occurs at $5 \leq t \leq 6$ or e_3 occurs at $8 \leq t \leq 10$. A fault is detected if none of the expected events occur within 6 time units after the occurrence of e_1, one of the expected events is observed out of their expected time interval, e.g. e_2 at time $t = 7$, or an unexpected event occurs, e.g. e_4. A situation in which this expectation is satisfied is illustrated in Figure 6.17. In this example, only one expectation is considered. In general, the set $O(t)$ can contain more than one expectation. The authors show that template models are as powerful as automata models to represent languages of DES. Furthermore, an extension to the so-called multiple-instance processes is available for considering systems with concurrency. Templates can model, besides the fault-free system behavior, also fault related information. In [Sayed-Mouchaweh, 2012], a set of expectations is generated based on an automaton model of the fault-free system behavior. The expectation definition is extended in order to contain information about the cause of faults in case of their violation. However, since the model provides no information about the faulty system behavior, it must be manually added by an expert.

6.4.4 Discussion

The introduced model-based fault diagnosis approaches automatically perform the fault detection and isolation tasks. In order to diagnose *time related faults*, such as deadlocks for instance, the model needs to contain information about the time behavior of the system. This is usually modeled in terms of time bounds that constrain the state transition, as presented by [Hashtrudi Zad et al., 2005], [Supavatanakul et al., 2006], [Jiroveanu and Boel, 2006], and [Pandalai and Holloway, 2000]. Many approaches for timed FDI result from the extension of an untimed approach, as for instance the standard diagnoser approach [Sampath et al., 1995] and the timed diagnoser [Hashtrudi Zad et al., 2005].

The presented model-based FDI approaches are *passively* operating, without affecting the operation of the system, and follow the idea of *consistency testing*. If *fault models* are used, fault diagnosis refers to checking whether the observed behavior is consistent with the modeled normal or with any faulty behavior. This concept is pursued in the diagnoser approach of [Sampath et al., 1995], in the ILP approach of [Dotoli et al., 2009], and in [Supavatanakul et al., 2006]. For systems for which fault information is not available, model-based FDI approaches using *fault-free models* have to be preferred, as discussed in Chapter 2. In [Pandalai and Holloway, 2000], fault detection is performed by checking the consistency between the observed and modeled fault-free behavior of the system. Fault-free models can also be used to isolate faults, as presented in [Reiter, 1987] and [Roth, 2010]. In these works, a set of faulty components is determined such that the observed inconsistencies between the observed and the modeled fault-free behavior can be explained. These approaches are *generic* since they deal with any type of fault that is a priori unknown. However, it cannot be guaranteed that a fault can always be isolated correctly. Since no information about the faulty system behavior is modeled, the isolation of faults has to rely on the interpretation of the observed fault symptoms. This interpretation is basically not unique, as discussed in [Reiter, 1987].

To minimize the number of missed and false detections, different concepts are proposed. In [Sampath et al., 1995], the *diagnosability* property is introduced to guarantee the detection and isolation of modeled faults. Since this requires the knowledge of faulty system behavior, a corresponding diagnosability property does not exist for approaches in which fault-free models are used. A possibility to improve fault diagnosis is to perform *decentralized* diagnosis, see [Debouk et al., 2000] and [Roth, 2010], and *distributed* diagnosis, see [Genc and Lafortune, 2003] and [Jiroveanu and Boel, 2006]. Decentralized and distributed FDI is especially advantageous if large and concurrent systems are considered. In general, the number of missed and false detections obtained with fault diagnosis mainly depends on the proper design of the system model.

For *online* fault diagnosis, the calculation efforts of the diagnosis algorithms have to be considered. The approaches of [Sampath et al., 1995], [Roth, 2010], and [Pandalai and Holloway, 2000] have basically low calculation demands since the approaches rely mainly on a state observation task that can be efficiently implemented. This has been shown in [Danancher et al., 2011] by means of a case study. In contrast to that, ILP problems, which have to be solved in [Dotoli et al., 2009], pose high computational complexity, especially if the systems are large and complex. This prevents to use ILP-based FDI method for real industrial systems. The *economic efficiency* of a model-based FDI implementation depends mainly on the required modeling efforts. Therefore, approaches are advantageous that apply diagnosis models that can be automatically be generated, as in [Roth, 2010], [Supavatanakul et al., 2006], and [Pandalai and Holloway, 2000], for instance.

Chapter 7

Conclusion

7.1 Summary

In this thesis, a framework was proposed that allows to perform model-based fault diagnosis for industrial DES based on system observations and a minimum amount of system knowledge. Therefore, new approaches for identification of timed FDI models, partitioning of FDI models, and timed fault diagnosis procedures were presented. It has been explained how fault diagnosis models are automatically determined based on observed timed system behavior and how these models can be used to successfully detect and isolate faults. The framework is especially appropriate for systems that are large and complex for which no knowledge about the behavior is available and in the case that a diagnosis implementation must meet economic constraints. This holds for many manufacturing and production applications in industry, especially for existing systems that are operating for dozens of years. Besides the theoretical explanations, which were presented for a formal and clear development of the methods, the thesis provided practically relevant results by the application to a real-world evaluation system.

A new timed identification algorithm has been presented that automatically identifies the behavior of a timed automaton. An identified timed automaton represents the timed fault-free behavior of a DES. The timed fault-free behavior is modeled to be able to detect and isolate faults based on time fault symptoms, such as the early, late, and deadlock behavior. Specific guarantees were given for precision and completeness of the identified models. This is important to ensure that the number of missed and false detections is minimal when using the identified models for FDI purposes. Most identification approaches from literature do not consider this particular issue. Applying the identification algorithm to the BMS resulted in a timed automaton that consists of thousands of states and transitions. This automaton can hardly be manually modeled, even if the required system knowledge would be available. For systems with concurrency, a timed distributed modeling approach and a distributed extension for the identification algorithm have been introduced. It has been discussed, how distributed modeling can be used to determine logically complete models for concurrent systems, even if a monolithic model cannot be completely identified based on the same given set of observed timed sequences. This has been demonstrated by means of the BMS. A further important result of the distributed BMS identification is that the number of model states and transitions could be significantly reduced, in comparison with the monolithic model, and that the data-basis for time identification was improved. In that way, useful time guards could be determined for almost all transitions of the distributed model. In order to model the dependent subsystems behavior of the DES, shared I/Os were considered. It has been shown, by means of an example, how such I/Os can contribute to the minimization of exceeding behavior in order to improve the fault detection capabilities of model-based fault diagnosis.

Before a distributed model can be identified, the model needs to be appropriately partitioned. Only few works in literature deal with the automatic partitioning of DES models. In this thesis, new data-based approaches were presented that rely on observed output sequences of the system and require only a minimum amount of system knowledge. The idea is first, to divide the global partitioning problem into several local problems, second, to solve the local problems, and third, to synthesize the I/O-partition for the DES out of the solutions of the local problems. An advantage of this procedure is that the computational complexity of partitioning can be distributed among several, smaller problems that can be solved in parallel. Furthermore, I/O-partitions for large and complex closed-loop DES with a high degree of concurrency can be determined and the resulting I/O-partitions contain shared I/Os. While causal partitioning determines causal relationship between I/Os in order to generate possible I/O-subsets, optimal partitioning relies on an optimization that is formulated as a Knapsack problem. An algorithm was presented that synthesizes valid I/O-partitions based on the determined I/O-subsets such that complete models with minimal exceeding behavior can be identified. The application to the BMS showed that causal partitioning may not find I/O-subsets for all I/Os of the closed-loop DES because of conditional behavior and noisy data. Optimal partitioning tended to be robust against these influences but was more challenging from the computational point of view. The finally automatically synthesized I/O-partitions for the BMS even outperformed an expert I/O-partition with respect to the language growth and the number of I/O-subsets.

Model-based concepts are often used in literature for fault diagnosis of DES. However, many existing works consider only academic examples that allow for manual modeling including the knowledge about faulty system behavior. The BMS model and especially models of real-word DES can consist of thousands of states and transitions while faults and their effects are typically unknown before they actually appear. In this context, a new timed FDI approach was proposed to detect and isolate faults by using a model of the fault-free timed DES behavior. This model can be identified using the approaches for automatic modeling that were presented in this thesis. The procedures for timed fault detection and isolation rely mainly on the timed estimation of the current model state using the observed behavior from the DES. This information is used for fault detection to decide about the fault state of the system. Detected faults can be isolated by the proposed DES specific deadlock, early, and late behavior residuals. The application to the BMS showed that the presented FDI approach is appropriate for detecting and isolating generic faults. A fault can basically be detected if the DES generates fault symptoms, i.e. behavior can be observed that differs from the fault-free one. By means of the residuals, a set of fault candidates can be determined, which can be used by the maintenance crew to isolate the faulty system component. The evaluation of time identification parameters showed the parameters influence on the fault detection capabilities and on the number of false detections. In comparison with a normal distribution approach from literature, the proposed relative tolerance approach was more appropriate for the time guard determination of BMS models since less false detections were made for comparable precise models. The capabilities of a distributed modeling approach were demonstrated in a second evaluation. It has been shown that the number of false detections can be significantly reduced by using a distributed modeling approach for the concurrent operating BMS. Furthermore, the results also showed that distributed modeling contributes to the fault detection capabilities. By avoiding false detection, the FDI implementation required less time for reinitialization and none of the studied example faults were missed.

7.2 Further Work

It has been shown in this thesis that the properties of identified time guards, i.e. precision and completeness, depend significantly on the data-base that is used for identification. Typically, the available data differs for each transition leading to different initial conditions for the identification. One way to improve identified time guards is to do *online updating* while applying the model. Therefore, appropriate algorithms need to be developed that continuously enlarge the time data-base of each transition by new observations and run time guard determination again based on the modified data sets. An especially interesting question in this context is how distributed models can be updated online when the partial timed behavior needs to be considered instead of the monolithic one.

The closed-loop DES considered in this work use exclusively sensors and actuators that have binary values. Another possible research topic for further work is to extend the presented automatic modeling and FDI approaches to systems that use signals with *continuous values*. In [Supavatanakul et al., 2003], an approach is proposed in which these signals are quantized such that their progress over time can be modeled by means of discrete events. It is worth to investigate how this procedure can be adopted for the presented FDI approach and how practical implementation issues with real-world systems, such as the parametrization of the quantization and the state-space explosion problem for instance, can be dealt with. Another question is whether hybrid models can be used in order to simplify the modeling procedure of these systems.

Partitioning of DES models has been performed in this thesis with respect to the logical behavior of the system. The partitioning aim was formulated with respect to the logical completeness of the resulting distributed model. It may be possible to further improve partitioning by incorporating *time information* contained in the observed data and by considering the completeness time the modeled time behavior. This can be done by the formulation of an alternative optimization criterion for optimal partitioning that focuses on the optimization of the logical model structure such that the data-base available for time identification is improved. Another possible extension of this may be multi-objective optimization in order to combine time and logically related cost functions.

Chapter 8

Kurzfassung in deutscher Sprache (extended summary in German)

In der Industrie leisten Fehlerdiagnoseverfahren einen wichtigen Beitrag zur Gewährleistung der Verfügbarkeit von Anlagen und Maschinen. Dabei stehen vor allem ungeplante Stillstandszeiten aufgrund defekter Anlagenkomponenten im Fokus, welche die Verfügbarkeit einer Anlage drastisch reduzieren können. Automatische Fehlerdiagnoseverfahren ermöglichen im Falle eines Fehlers die zeitnahe Erkennung und Isolierung von defekten Anlagenkomponenten. Dadurch wird eine in der Regel zeitaufwändige, manuelle Fehlersuche vermieden und erforderliche Instandsetzungsmaßnahmen können unmittelbar eingeleitet werden. In dieser Arbeit wurde ein modellbasiertes Fehlerdiagnoseverfahren für gesteuerte, industrielle Systeme entwickelt. Modellbasierte Fehlerdiagnoseverfahren basieren auf dem Prinzip der Konsistenzprüfung von beobachtetem System- und Modellverhalten. Auf dieser Basis kann entschieden werden, ob das beobachtete Systemverhalten einem Fehlverhalten entspricht und welche Komponenten des Systems gegebenenfalls dafür verantwortlich sein können. Zur Anwendung modellbasierter Fehlerdiagnose ist es erforderlich, ein geeignetes Modell des industriellen Systems zu entwickeln. Diese Systeme bestehen in der Regel aus vielen, miteinander interagierenden Komponenten und realisieren komplexe, zeitabhängige Prozesse mit potentiell nebenläufigem Verhalten. Dabei ist die Dokumentation von Hard- und Software sowie das Wissen von Experten über den Prozess und mögliche Fehlerursachen oftmals unvollständig oder veraltet. Zur wirtschaftlichen Modellierung solcher Systeme sind Verfahren von Vorteil, die automatisch und datenbasiert arbeiten und weitestgehend ohne Expertenwissen auskommen. Hierfür wurden in dieser Arbeit neue Methoden zur zeitbehafteten Modellidentifikation und Modellpartitionierung entwickelt, die ausschließlich auf Basis des beobachteten, fehlerfreien Systemverhaltens arbeiten. Neben den erforderlichen theoretischen Studien wurden ebenso die zugehörigen praktischen Aspekte der Methoden, anhand der exemplarischen Implementierung am Bosch Mechatronics System (BMS), eingehend untersucht.

Die in dieser Arbeit betrachteten industriellen Systeme bestehen aus einem geschlossen Kreis von Steuerung und Anlage, wobei die zugehörigen Sensoren und Aktoren einen ausschließlich binären Wertebereich haben. Ein solches System, wie in Abbildung 8.1 gezeigt, wurde als ein ereignisdiskretes System modelliert, d.h. der Zustand des Systems hängt

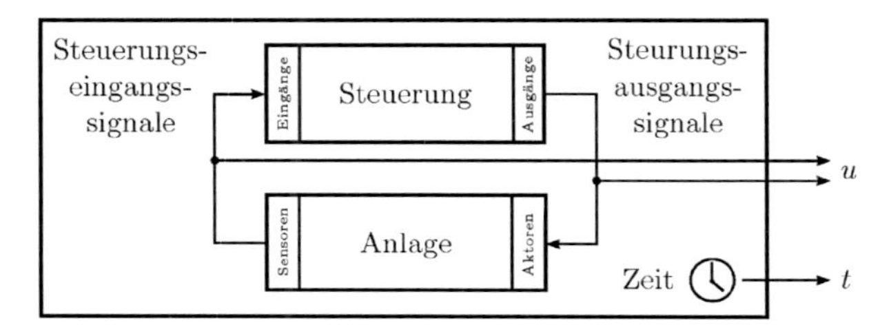

Abbildung 8.1: Industrielles ereignisdiskretes System

ausschließlich vom Auftreten diskreter Ereignisse über der Zeit ab. Aufgrund des geschlossenen Kreises von Steuerung und Anlage wird ein solches System auch als 'closed-loop' bezeichnet. An die Fehlerdiagnose eines industriellen ereignisdiskreten Systems werden bestimmte Anforderungen gestellt. Die Fehlerdiagnose muss in der Lage sein Fehlverhalten des Systems automatisch zu erkennen und die betreffenden defekten Komponenten in der Anlage zu isolieren. Dabei sollen auch Fehler diagnostiziert werden können, die sich durch zeitliches Fehlerverhalten bemerkbar machen. Eine weitere, wesentliche Anforderung ist die Minimierung von nicht-erkennbaren Fehlern und Falscherkennungen. Aus diesen Anforderungen lassen sich wiederum Schlussfolgerungen für ein Modell und dessen Entwicklung ableiten. Für zeitbehaftete Fehlerdiagnose muss ein verwendetes Modell das Zeitverhalten des ereignisdiskreten Systems abbilden. Aufgrund des daten-basierten Modellierungsansatzes und des unvollständigen Wissens liegt es nahe Modelle zu verwenden, die ausschließlich das fehlerfreie, zeitbehaftete Anlagenverhalten repräsentieren. Fehlerfreies Verhalten bedeutet dabei, dass ein bestimmtes Anlagenverhalten von einem verantwortlichen Anlagenfahrer akzeptiert wurde. Zur Minimierung von nicht-erkennbaren Fehlern und Falscherkennungen ist es notwendig, dass ein verwendetes Modell präzise und vollständig ist. Präzise Modelle modellieren ausschließlich Verhalten, welches von dem originalen System wiedergeben werden kann. Vollständigkeit bedeutet, dass Modelle in der Lage sind das gesamte originale Verhalten des betrachteten Systems wiederzugeben. Da jedoch das originale Systemverhalten unbekannt ist, kann im Allgemeinen nicht garantiert werden, dass ein automatisch erstelltes Modell gänzlich präzise und vollständig ist. Das Ziel der automatischen Modellierung ist es vielmehr Modelle zu generieren, welche die genannten Anforderungen bestmöglich hinsichtlich der zur Verfügung stehenden Beobachtungen erfüllen.

Zur Beobachtung des ereignisdiskreten Systems wurden die Werte der Sensor- und Aktuatorsignale, welche die beobachtbaren Systemausgabe u darstellen, sowie der zugehörigen Zeitverlauf t erfasst. Bei dem verwendeten Modell handelt es sich um einen zeitbewerteten autonomen Automaten mit Ausgabe (TAAO). Dieser Zeitautomat ist in der Lage das logische, d.h. die Systemausgabe des ereignisdiskreten Systems, und das zugehörige zeitliche Verhalten zu modellieren. Zur Beschreibung des zeitbewerteten Verhaltens von System und Modell wurde eine zeitbewertete Sprache L_t, eine logische Sprache L, sowie Mengen und Folgen von relativen Zeiten Δ eingeführt. Sie ermöglichen die differenzierte Betrachtung der logischen und zeitlichen Verhaltensdimensionen und die getrennte Identifikation des logischen und zeitlichen Modellverhaltens. Für die Identifikation des logischen Modellverhaltens wurde ein existierender Algorithmus nach [Roth, 2010] verwendet. Automaten, identifiziert nach diesem Algorithmus, sind logisch präzise und vollständig hinsichtlich eines Identifikationsparameters k. Das zeitliche Verhalten des Modells wird mittels einer neuen Methode auf Basis von nicht-parametrischen Toleranzintervallen gewonnen. In Abhängigkeit von der Anzahl verfügbarer Beobachtungen und der Wahl des Identifikationsparameters ext_0 kann ein identifizierter Automat zeitlich präzise identifiziert werden. Das bedeutet, dass die ermittelten Zeitintervalle, die das Zeitverhalten jeder Transition beschränken, so gewählt sind, dass keine relativen Zeiten modelliert werden die kleiner oder größer als die beobachteten relativen Zeiten sind. Weiterhin kann die Vollständigkeit des modellierten Zeitverhaltens eines identifizierten Automaten gezeigt werden, unter der Bedingung, dass die originale logische Sprache des ereignisdiskreten Systems vollständig beobachtet wurde und die Grenzen der originalen Zeiten den beobachteten Grenzen entsprechen. Im Allgemeinen sind die Präzision und die Vollständigkeit eines Modells gegenläufige Eigenschaften. Das Ziel der Modellierung ist es, durch eine geeignete Wahl der Identifikationsparameter einen gewünschten Kompromiss, hinsichtlich Modellgenauigkeit und -vollständigkeit, zu erreichen. Hierzu wurden in dieser Arbeit die Eigenschaften der Identifikationsparameter und deren Einflüsse näher beleuchtet.

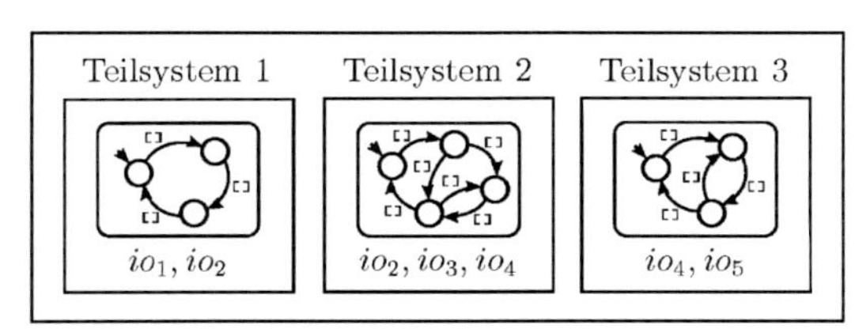

Abbildung 8.2: Nebenläufiges ereignisdiskretes System

Viele industrielle Systeme setzen sich aus einer Anzahl von nebenläufigen Teilsystemen zusammen. Dabei repräsentiert ein Teilsystem jeweils das Verhalten einer Teilmenge aller Steuerungsein- und Steuerungsausgänge (I/Os) des ereignisdiskreten Systems. Ein Beispiel für solch ein nebenläufiges System ist in Abbildung 8.2 gezeigt. Das Verhalten jedes Teilsystems lässt sich auf einen individuellen Zustandsautomaten zurückführen, wobei das Verhalten der Automaten verkoppelt sein kann. Nebenläufigkeiten und Verkopplungen lassen sich anhand der I/Os erkennen, die dem jeweiligen Teilsystemen zugeordnet sind. Zur vollständigen Identifikation eines monolithischen Modells ist es erforderlich, das originale Verhalten des Systems vollständig zu beobachten. Bei Systemen mit Nebenläufigkeit kann dies eine inakzeptabel lange Beobachtungszeit zur Folge haben. Die Ursache hierfür liegt in der großen Zahl an Möglichkeiten, die sich aus dem kombinierten Verhalten der einzelnen Teilsysteme ergeben können. Zur vollständigen Identifikation solcher Systeme, auf Basis einer gegebenen Menge an Beobachtungen, wurde in dieser Arbeit eine verteilte Modellierung verwendet. Hierdurch wurde es ermöglicht, vollständige Modelle für Systeme zu identifizieren, für die das monolithische Modell, identifiziert auf Basis der gleichen Daten, unvollständig war. Dies wurde in der Arbeit am Beispiel des BMS gezeigt. Weiterhin konnte durch die verteilte Modellierung die Anzahl der Zustände und Transitionen im Modell verringert und die Datenbasis für die zeitbehaftete Identifikation verbessert werden, so dass mehr Zeitbeschränkungen in resultierenden Automaten ermittelt wurden. Der Nachteil einer solchen Modellierung liegt in der verringerten Präzision des identifizierten Modells. Dem Präzisionsverlust kann zwar durch geeignete Maßnahmen entgegengewirkt werden, z.B. durch eine minimale Anzahl an Teilmodellen und durch gemeinsame I/Os, jedoch können keine Garantien für die Modellpräzision mehr gegeben werden. Der Einfluss von gemeinsam modellierten I/Os wurde in der Arbeit anhand eines Beispielmodells gezeigt.

Zur Identifikation eines verteilten Modells ist es zunächst erforderlich, das Modell hinsichtlich der Nebenläufigkeit geeignet zu partitionieren. Nur wenige Arbeiten in der Literatur beschäftigen sich mit dieser Aufgabenstellung. In dieser Arbeit wurden hierfür Methoden vorgestellt, die automatisch und datenbasiert die verteilte Struktur eines ereignisdiskreten Systems ermitteln, ohne auf zusätzliche Dokumentation oder Expertenwissen zurückgreifen zu müssen. Den Methoden liegt die Idee zugrunde, das globale Partitionierungsproblem in mehrere Teilprobleme aufzuteilen, diese zu lösen, und anschließend die Partitionierung des Systems anhand der ermittelten, lokalen Lösungen zu bestimmen. Die Vorteile dieser Herangehensweise liegen in der Möglichkeit Partitionen auch für große Systeme mit einem hohen Grad an Nebenläufigkeit zu bestimmen, damit Berechnungen parallelisiert und verhaltensgekoppelte Teilmodelle über gemeinsame I/Os generiert werden können. Die drei Verfahren, die in dieser Arbeit zur Partitionierung vorgestellt wurden, sind die kausale und optimale Partitionierung, welche die lokalen Partitionierungsprobleme lösen und die Partitionssynthese, welche die globale Partitionierung auf Basis der lokalen Lösungen ermittelt. Bei der kausalen Partitionierung werden Mengen von I/Os bestimmt, die in einer kausalen Beziehung zueinander stehen. Kausale Gruppen repräsentieren Teilsysteme die sequentiell arbeiten und deren Verhalten in jedem Produktionszyklus gleich ist. Solche Teilsysteme können auf Basis der beobachteten Daten in der Regel vollständig identifiziert werden. Bei der optimalen Partitionierung wurde die Bestimmung geeigneter I/O-Mengen als ein Optimierungsproblem formuliert. Es werden solche I/O-Mengen gesucht die maximal groß sind und die auf Basis der zur Verfügung stehenden Daten als vollständig identifizierbar betrachtet werden können. Die maximale Größe soll ermöglichen, dass für die Partitionssynthese möglichst wenige Teilmodelle benötigt werden und somit die Modellpräzision maximiert wird. Die Partition des Modells wird über die minimale Anzahl aller ermittelten I/O-Mengen synthetisiert, so dass alle I/Os des ereignisdiskreten Systems beinhaltet sind. Die Anwendung am BMS zeigte, dass nicht für alle I/Os immer andere kausale I/Os gefunden werden können. Dies motiviert die Anwendung der optimalen Partitionierung. Das Verfahren lieferte auch für nicht-kausale I/Os mögliche I/O-Mengen, die für die Partitionssynthese verwendet werden können, jedoch erforderte dies auch einen höheren Rechenaufwand als die kausale Partitionierung. Die synthetisierten Partitionen wurden mit Expertenlösungen verglichen. Dabei zeigte sich, dass die automatisch generierten Lösungen die Expertenlösungen hinsichtlich der geschätzten Vollständigkeit sogar übertreffen können. Zusätzlich enthalten die automatisch ermittelten Partitionen gemeinsame I/Os zur Erhöhung der Modellpräzision, was bei dem verteilten Expertenmodell nicht der Fall war.

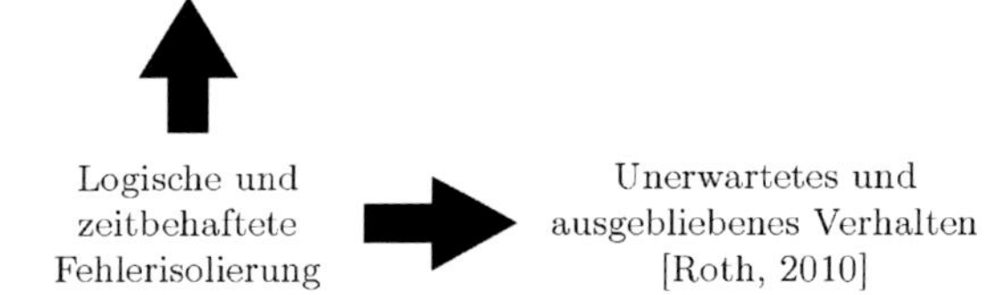

Abbildung 8.3: Dimensionen der Fehlerisolierung

Modellbasierte Fehlerdiagnose ist ein in der Forschung und Anwendung weit verbreiteter Ansatz zur Erkennung und Isolierung von Fehlern. Verfahren aus der Literatur gehen oftmals davon aus, dass die notwendigen Modelle händisch erstellt werden können und das Fehlverhalten der Anlage vollständig bekannt ist. Für industrielle Systeme ist dies typischerweise nicht der Fall. Weiterhin konnte in dieser Arbeit gezeigt werden, dass Modelle von realen Systemen durchaus tausende von Zuständen und Transitionen beinhalten können, die händisch nicht wirtschaftlich zu bestimmen sind. In diesem Kontext wurde ein zeitbehaftetes Fehlerdiagnoseverfahren präsentiert, dass die Fehlererkennung und -isolierung anhand eines fehlerfreien, zeitbehafteten Modells durchführt. Das erforderliche Modell kann mittels der vorgestellten Identifikations- und Partitionierungsverfahren automatisch gewonnen werden. Das Prinzip zur Fehlererkennung und -isolierung basiert im Wesentlichen auf der zeitbehafteten Schätzung des aktuellen Modellzustandes. Unter Verwendung dieser Information können Fehler erkannt und mittels DES spezifischer Residuen isoliert werden. Die Fehlerisolierung erfolgt, wie schon die Identifikation, auf der getrennten Betrachtung von logischem und zeitlichem Verhalten, wie in Abbildung 8.3 gezeigt. In der logischen Dimension werden Fehlersymptome im Hinblick auf unerwartetes und ausgebliebenes Verhalten betrachtet. Hierfür wurden in [Roth, 2010] geeignete Residuen vorgestellt. Die neu entwickelten Residuen der zeitlichen Dimension berücksichtigen die Fehlersymptome blockierendes, frühes und spätes Verhalten. Alle Residuen liefern nach ihrer Berechnung eine Teilmenge der I/Os des ereignisdiskreten Systems, welches die möglichen Fehlerkandidaten enthält. Da die Menge der Fehlerkandidaten in der Regel klein ist, kann der zugrunde liegende Fehler in der Anlage zeitnah bestimmt werden. Dies wurde durch die Anwendung des Fehlerdiagnoseverfahrens am BMS bestätigt. Die betrachteten Beispielfehler wurden nach ihrem Auftreten unmittelbar erkannt und isoliert. Dabei waren die fehlerhaften Anlagenkomponenten in allen Fällen in den Kandidatenlisten enthalten. Die Evaluierung der Parameter für die zeitbehaftete Identifikation, in Hinblick auf die Fehlerdiagnose, zeigte den unmittelbaren Zusammenhang zwischen der Erhöhung der Toleranz bei der Zeitidentifikation und der damit verbunden Reduktion von Falscherkennungen. Dabei ermöglichte der entwickelte Zeitidentifikationsansatz mittels Toleranzintervallen eine deutlichere Verringerung der Falscherkennungen bei vergleichbarer Toleranz, als ein Ansatz der auf der Annahme normalverteilter Zeiten beruht. Durch die Evaluierung des verteilten Modellansatzes wurde bestätigt, dass unter Verwendung eines solchen Modells bei nebenläufigem Verhalten die Anzahl logischer und zeitlicher Falscherkennungen deutlich reduziert werden kann. Weiterhin wurde gezeigt, dass die verteilte Modellierung auch deutlich zur Reduzierung von nicht-erkennbaren Fehlern beiträgt. Dies lässt sich dadurch erklären, dass durch die geringere Anzahl an Falscherkennungen weniger Zeit für Reinitialisierungen des Diagnosesystems erforderlich und somit die Fehlererkennung höher verfügbar war. Bei der Anwendung auf das BMS wurden mit einem verteilten Modell sämtliche Beispielfehler erkannt, wobei hingegen mit dem monolithischen Modell nur Einer von Vieren erkannt werden konnte.

Appendix A

Proofs

Lemma 1. $\forall w^q \in L_{Obs}^{k+1,\Sigma_t^k}$ with $\Delta_{Obs,w^q[q]}^{w^q} \neq \perp$, $\exists w^{k+1} \in L_{Obs}^{k+1,\Sigma_t^k}$ such that $w^q = w^{k+1}\langle k - q + 2..k+1\rangle$.

Proof of Lemma 1. The proofs will show that for any word $w^q \in L_{Obs}^{k+1,\Sigma_t^k}$ of length $q \leq k+1$ with time spans observed for the q-th symbol a word $w^{k+1} \in L_{Obs}^{k+1,\Sigma_t^k}$ of length $k+1$ exists, such that the last q symbols of w^{k+1} correspond to w^q. In other words, the word w^{k+1} ends with the symbols given by w^q. From Definition 16, L_{Obs}^{k+1,Σ_t^k} is determined. Therefore, observed timed output sequences Σ_t are $k-1$-time duplicated by Equation 3.25. Following Assumption 1, $\delta_h(1) = \perp$ holds $\forall \sigma_{t,h} \in \Sigma_t$. Due to the duplication of $\delta_h(1) = \perp$, $\Delta_{Obs,w^q[q]}^{w^q} = \perp$ holds $\forall w^q$ with $q < k+1$ based on the first k symbols $w^q = (u_{t,h}(1), u_{t,h}(2), \ldots, u_{t,h}(q))$ of the h-th observed sequence of Σ_t^k. For these words, denoted as $\hat{L}_{Obs}^{k+1,\Sigma_t^k}$, $\nexists w^{k+1} \in L_{Obs}^{k+1,\Sigma_t^k}$ such that $w^q = w^{k+1}\langle k-q+2..k+1\rangle$, since for all $k-1$ words $w^{k+1} = (u(1), u(2), \ldots, u(k+1))$, determined by Equation 3.8, $u(k+1) \neq u_{t,h}(q)$, $q < k+1$. For all other words $w^q \in L_{Obs}^{k+1,\Sigma_t^k} \setminus \hat{L}_{Obs}^{k+1,\Sigma_t^k}$, $\Delta_{Obs,w^q[q]}^{w^q} \neq \perp$ since $\delta(j) \neq \perp, j > 1$, according to Definition 8 and due to the construction rule for words with length $k+1$ after Equation 3.8. The construction of words w^{k+1} start with $w^{k+1} = (u(1), u(2), \ldots, u(k+1))$ and iterates over each $\sigma_{t,h}$, such that for each words w^q with $w^q \in L_{Obs}^{k+1,\Sigma_t^k} \setminus \hat{L}_{Obs}^{k+1,\Sigma_t^k}$ at least once $w^q = w^{k+1}\langle k-q+2..k+1\rangle$ holds. □

Theorem 1. If a *TAAO* is logically precise and timed identified such that $|\Delta_{Obs}^{(x,x')}| \geq v_0$ $\forall (x, guard, c, x') \in TT$ or $ext_0 = 0$, then $\forall w^q \in L_{Mod}^{k+1}$, the q-th identified time span set $\Delta_{Mod,w^q[q]}^{w^q}$ of $\Delta_{Mod}^{w^q} = (\Delta_{Mod,w^q[1]}^{w^q}, \Delta_{Mod,w^q[2]}^{w^q}, \ldots, \Delta_{Mod,w^q[q]}^{w^q})$ is *precisely bounded*, such that $\min\left(\Delta_{Mod,w^q[q]}^{w^q}\right) = \min\left(\Delta_{Obs,w^q[q]}^{w^q}\right)$ and $\max\left(\Delta_{Mod,w^q[q]}^{w^q}\right) = \max\left(\Delta_{Obs,w^q[q]}^{w^q}\right)$ holds.

Proof of Theorem 1. It is proven in the following that the q-th modeled time bounds of an identified *TAAO* are precisely bounded if the *TAAO* is logically precise and if $|\Delta_{Obs}^{(x,x')}| \geq v_0$ $\forall (x, guard, c, x') \in TT$ or $ext_0 = 0$ holds. The argumentation is based on the automaton before the output is determined in Algorithm 2. First, the theorem is proven for words with $q = k+1$ and then it is generalized for words with $q < k+1$. Given the logical precision, it follows that for each $w^q \in L_{Obs}^{k+1}$ exactly one word $w^q \in L_{Mod}^{k+1}$

exists, $1 \leq q \leq k+1$. The allocation of observed time spans according to Algorithm 2 is performed for words $w^{k+1} \in W_{Obs}^{k+1,\Sigma_t^k}$ and the related time span collections $\Delta_{Obs}^{w^{k+1},\Sigma_t^k}$. Since only the $(k+1)$-th time span set of $\Delta_{Obs}^{w^{k+1},\Sigma_t^k}$ is considered, the k-times artificially extension at the beginning of a word after Equation 3.25 has no influence on the applied time spans. In Line 8, the algorithm sets $\Delta_{Obs}^{(x,x')} := \Delta_{Obs,w^{k+1}[k+1]}^{w^{k+1},\Sigma_t^k}$ with $\lambda(x) = w^{k+1}\langle 1..k\rangle$ and $\lambda(x') = w^{k+1}\langle 2..k+1\rangle$. The time guards of these transitions are determined by Algorithm 3 according to the premises: If either $|\Delta_{Obs}^{(x,x')}| \geq v_0\ \forall (x, guard, c, x') \in TT$ or $ext_0 = 0$, then $guard(x,x') = \left[\min\left(\Delta_{Obs}^{(x,x')}\right), \max\left(\Delta_{Obs}^{(x,x')}\right)\right]$. As a result, for each word w^{k+1} exactly one transition is modeled with $guard(x,x')$ based on $\Delta_{Obs,w^{k+1}[k+1]}^{w^{k+1},\Sigma_t^k}$. With Equation 3.19, the modeled time spans of w^{k+1} are $\Delta_{Mod}^{w^{k+1}} = (\Delta_{Mod,w^{k+1}[1]}^{w^{k+1}}, \Delta_{Mod,w^{k+1}[2]}^{w^{k+1}}, \ldots, \Delta_{Mod,w^{k+1}[k+1]}^{w^{k+1}})$. $\Delta_{Mod,w^{k+1}[k+1]}^{w^{k+1}}$ are the time spans of the time guard $guard(x,x')$ for a transition for which the same condition holds as before: $\lambda(x) = w^{k+1}\langle 1..k\rangle$ and $(\lambda(x') = w^{k+1}\langle 2..k+1\rangle$. As a result,

$$\min\left(\Delta_{Mod,w^{k+1}[k+1]}^{w^{k+1}}\right) = \min\left(guard(x,x')\right) = \min\left(\Delta_{Obs,w^{k+1}[k+1]}^{w^{k+1}}\right)$$

and

$$\max\left(\Delta_{Mod,w^{k+1}[k+1]}^{w^{k+1}}\right) = \max\left(guard(x,x')\right) = \max\left(\Delta_{Obs,w^{k+1}[k+1]}^{w^{k+1}}\right).$$

The next step is to extend the proof for words $w^q \in L_{Obs}^{k+1}$ with $q < k+1$. Two cases have to be considered: $w^q \in L_{Obs}^{k+1,\Sigma_t^k}$ with $\Delta_{Obs,w^q[q]}^{w^q} = \bot$ and $w^q \in L_{Obs}^{k+1,\Sigma_t^k}$ with $\Delta_{Obs,w^q[q]}^{w^q} \neq \bot$. The first case represents words w^q with undefined time spans for the q-th symbol. No time guard will be identified based on these words, hence they are ignored in the following. For words of the second case, the following holds: Since more than one word w^{k+1} may exist for w^q that is in accordance with Lemma 1, $\Delta_{Obs,w^q[q]}^{w^q} \supseteq \Delta_{Obs,w^{k+1}[k+1]}^{w^{k+1}}$ holds in general. Applying the identification for all these w^{k+1} with $\Delta_{Obs,w^{k+1}[k+1]}^{w^{k+1}}$ leads then to a set of i time guards $guard_i(x,x')$ where $(\lambda(x) = w^{k+1}\langle 1..k\rangle) \wedge (\lambda(x') = w^{k+1}\langle 2..k+1\rangle)$. Consequently, the overall minimum and respectively maximum of all time guards $guard_i(x,x')$ must be considered resulting to

$$\min\left(\Delta_{Mod,w^q[q]}^{w^q}\right) = \min_i\left(guard_i(x,x')\right) = \min\left(\Delta_{Obs,w^q[q]}^{w^q}\right)$$

and

$$\max\left(\Delta_{Mod,w^q[q]}^{w^q}\right) = \max_i\left(guard_i(x,x')\right) = \max\left(\Delta_{Obs,w^q[q]}^{w^q}\right).$$

□

Theorem 2. If $L_{Orig}^{k+1} = L_{Obs}^{k+1}$ and $\forall w^q \in L_{Obs}^{k+1}$, $\min\left(\Delta_{Orig,w^q[i]}^{w^q}\right) = \min\left(\Delta_{Obs,w^q[i]}^{w^q}\right)$ and $\max\left(\Delta_{Orig,w^q[i]}^{w^q}\right) = \max\left(\Delta_{Obs,w^q[i]}^{w^q}\right)$ $\forall 1 \leq i \leq q$ with $\bot \notin \Delta_{Orig,w^q[i]}^{w^q}$, then $L_{Mod,t}^n \supseteq L_{Orig,t}^n$ with $n \geq 1$ for an identified *TAAO*.

Proof of Theorem 2. It is proven here that an identified *TAAO* simulates the original timed language $L_{Orig,t}^n$ of the DES if the *TAAO* is logically precise and the original time span sets for all i symbols and all words $w^q \in L_{Obs}^{k+1}$ are completely observed. For the proof, the logical completeness and time completeness is considered separately. A *TAAO* is complete if it is logically complete $L_{Mod}^n \supseteq L_{Orig}^n$ and temporally complete such that $\Delta_{Mod,w^q[i]}^{w^q} \supseteq \Delta_{Orig,w^q[i]}^{w^q}$ for all words $w^q \in L_{Mod}^n$ with $q \geq 1$. The logical completeness of an identified *TAAO* has been proven in [Roth, 2010]. In the remaining, the time completeness is proven. It is shown that the modeled time span sets $\Delta_{Mod,w^q[i]}^{w^q}$ for all i symbols of a words $\forall w^q \in L_{Mod}^n$ are complete because they are identified by means of the observed time sets $\Delta_{Obs,k+1}^{w^{k+1},\Sigma_t^k}$ that are complete according the given premises. Since logical precision and $L_{Mod}^{k+n} \supseteq L_{Orig}^{k+n}$ with $n \geq 1$ holds, for each word $w^q \in L_{Orig}^n$, $\exists w^q \in L_{Mod}^n$. $\forall w^q \in L_{Mod}^n$, the *TAAO* contains at least one state trajectory $x(j), x(j+1), \dots, x(j+q+k-1)$ of length $q+k$ such that $w^q[1] = \lambda(x(j+k-1)), w^q[2] = \lambda(x(j+k)), \dots, w^q[q] = \lambda(x(j+q+k-1))$. The state trajectories end with states that produce w^q as outputs. According to Algorithm 2 the transitions of each state trajectory are identified based on observed words $w^{k+1} \in W_{Obs}^{k+1,\Sigma_t^k}$ of length $k+1$. Each transition ending in the i-th symbol of w^q with $w^q[i] = \lambda(x')$ belongs to exactly one word $w^{k+1} \in L_{Obs}^{k+1,\Sigma_t^k}$ with the related time span sequences $\Delta_{Obs,k+1}^{w^{k+1},\Sigma_t^k}$ that is used to identify the time guard. Since the bounds of $\Delta_{Obs,w^q[i]}^{w^q}$ can be considered as converged for all symbols, according to the premises, the bounds of $\Delta_{Obs,w^{k+1}}^{w^{k+1},\Sigma_t^k}$ are also converged. The time guards are determined according to Algorithm 3 as intervals enclosing all the observed time span bounds. As result, the modeled time span sets $\Delta_{Mod,w^q[i]}^{w^q}$ for all i symbols result from the identification of $\Delta_{Obs,k+1}^{w^{k+1},\Sigma_t^k}$ with converged bounds, $\forall w^{k+1} \in L_{Obs}^{k+1,\Sigma_t^k}$ that are part of the state trajectory and ending in w^q. Hence, $\Delta_{Mod,w^q[i]}^{w^q} \supseteq \Delta_{Orig,w^q[i]}^{w^q}$ for a word w^q with $q \geq 1$. □

Theorem 3. Given a *TDM* that is timed identified according to the I/O-partition $P = \{SUB_1, SUB_2, \dots, SUB_N\}$. If $L_{Mod,SUB_i,t}^n \supseteq L_{Orig,SUB_i,t}^n \ \forall 1 \leq i \leq N$, then $L_{Mod,TAAO_{\|},t}^n \supseteq L_{Orig,t}^n$ for $n \geq 1$.

Proof of Theorem 3. It is proven in the following that the composition of N partial automata of a *TDM* simulates the original timed language $L_{Orig,t}^n$ of a DES if all partial automata $TAAO \in TDM$ are completely identified. For the proof, logical and time completeness is separately considered. Logical completeness $L_{Mod,TAAO_{\|}}^n \supseteq L_{Orig}^n$ for $n \geq 1$ has been proven in [Roth, 2010]. The author showed that each logical word of the original logical language $w^q \in L_{Orig}^n$ can be reproduced by $L_{Mod,TAAO_{\|}}^n$. The proof for time completeness is given in the following. Therefore, a DES with two subsystems SUB_1 and SUB_2 is considered. A timed word $w_t^q \in L_{Orig,t}^n$ can be logically reproduced by the combined model $w^q = untime(w_t^q) \in L_{Mod,TAAO_{\|}}^n$ due to the logical completeness. There exists a state trajectory in $TAAO_{\|}$ that models the logical behavior of w^q. According to Equation 24, a timed word w_t^q can be temporally projected to two partial timed words $w_t^{q_1 \leq q} \in L_{Orig,SUB_1,t}^n$ and $w_t^{q_2 \leq q} \in L_{Orig,SUB_2,t}^n$. From the timed completeness precondition of the partial automata it follows that each partial model contains a state trajectory that reproduces the partial timed words $w_t^{q_1 \leq q}$ and $w_t^{q_2 \leq q}$, respectively. These local state trajectories are combined according to Definition 27 to a state trajectory in $TAAO_{\|}$. The related transitions are generated by Equation 3.41 such that the time guards of the resulting state trajectories w_t^q are consistent with the time guards in the partial models related to $w_t^{q_1 \leq q}$ and $w_t^{q_2 \leq q}$. Therefore, the time guards $guard_{c_i}$ and clock resets $c_i := 0$ of the partial models are related to the corresponding transitions in $TAAO_{\parallel}$. Since for all $w_t^q \in L_{Orig,t}^n$ a state trajectory exists in $TAAO_1$ and $TAAO_2$, the composed model $TAAO_{\parallel}$ contains a state trajectory that models w_t^q. If this is extended to all N partial automata, the resulting composed automaton is able to reproduce the original timed language $L_{Orig,t}^n$ of a DES. □

Theorem 4. Given an I/O-subset SUB, if $d(io_i, io_j) = 0 \; \forall io_i, io_j \in SUB$, then $d(SUB) = 0$.

Proof of Theorem 4. If $d(io_i, io_j) = 0$ holds for all I/O-pairs $(io_i, io_j), io_i, io_j \in SUB$, then the observed partial output sequences are equal such that $equal(\sigma_{\{(io_i,io_j)\},g}, \sigma_{\{(io_i,io_j)\},h}) = true$ holds, $\forall \sigma_{\{(io_i,io_j)\},g}, \sigma_{\{(io_i,io_j)\},h} \in \Sigma_{\{(io_i,io_j)\}}$, respectively. $\Sigma_{\{(io_i,io_j)\}}$ represents one class of sequences $\Sigma_{\{(io_i,io_j)\},class}$. This holds $\forall io_i, io_j \in SUB$. Since the observed partial sequences $\Sigma_{\{(io_i,io_j)\}}$ are available for all I/O-pairs (io_i, io_j) that can be generated for SUB, Σ_{SUB} can be unambiguously reconstructed. Due to the equality of the projected sequences and the unambiguous reconstruction, $equal(\sigma_{SUB,g}, \sigma_{SUB,h}) = true$ holds $\forall \sigma_{SUB,g}, \sigma_{SUB,h} \in \Sigma_{SUB}$. Hence, Σ_{SUB} also represents one class of sequences resulting in $d(SUB) = 0$. □

Proposition 1. If a node of the search tree, given by SUB, represents a non-causal I/O-subset, i.e. $d(SUB) > 0$, then all child nodes SUB' of SUB are non-causal I/O-subsets, i.e. $d(SUB') > 0$.

Proof of Proposition 1. Given a $SUB \subseteq IO$. If SUB is a non-causal I/O-subset, then $\exists (io_i, io_j)$ such that $d(io_i, io_j) \neq 0$ with $io_i, io_j \in SUB$. When SUB is extended by $io_e \in IO \backslash SUB$ such that $SUB' = SUB \cup io_e$ follows, then $io_i, io_j \in SUB'$ and $d(io_i, io_j) \neq 0$ still holds. According to Theorem 4, $d(io_i, io_j) = 0 \; \forall io_i, io_j \in SUB'$ such that $d(SUB') = 0$. Since this does not hold $d(io_i, io_j) \neq 0$, it follows that $d(SUB') \neq 0$. Hence, all following I/O-subsets SUB with $\exists (io_i, io_j)$ such that $d(io_i, io_j) \neq 0$ represent non-causal I/O-subsets. □

Appendix B

Example Shared I/Os

Given the partial timed automata $TAAO_1$ and $TAAO_2$ in Figure 3.17 that are based on the I/O-partition $P_{disj} = \{\{S_1, S_2\}, \{S_3, A\}\}$ without shared I/Os. The composed automaton $TAAO_{\|} = TAAO_1 \parallel TAAO_2$, according to Definition 27, is depicted in Figure B.1.

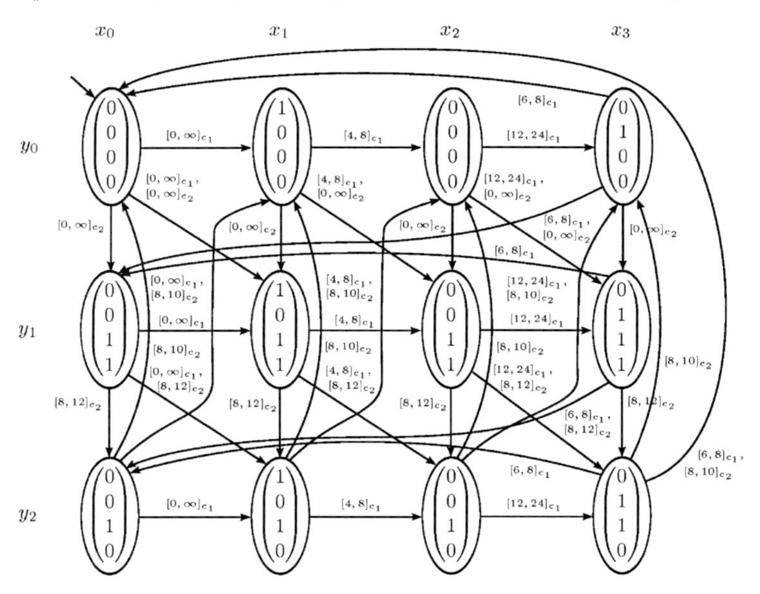

Figure B.1: Composed automaton $TAAO_{\|}$ based on $TAAO_1$ and $TAAO_2$ for P_{disj}

Given the partial timed automata $TAAO_1$ and $\widehat{TAAO_2}$ in Figure 3.19 that are based on the I/O-partition $P_{share} = \{\{S_1, S_2\}, \{S_2, S_3, A\}\}$ with the shared I/O S_2. The composed automaton $TAAO_{\|} = TAAO_1 \parallel \widehat{TAAO_2}$, according to Definition 27, is depicted in Figure B.2.

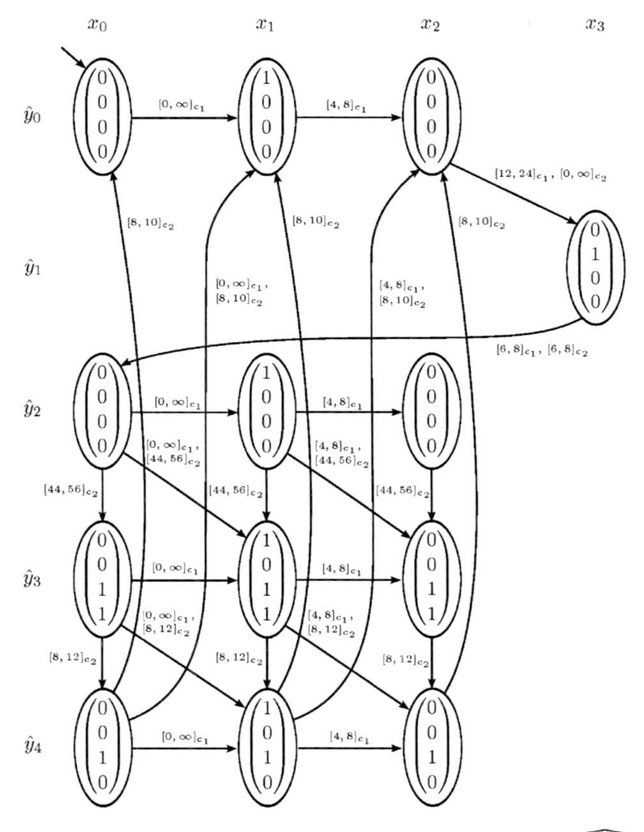

Figure B.2: Composed automaton $TAAO_{\|}$ based on $TAAO_1$ and $\widehat{TAAO_2}$ for P_{share}

Nomenclature

Acronyms

BMS Bosch Mechatronics System, page 4

DES Discrete Event System, page 2

FDI Fault detection and isolation, page 1

I/O Controller input and output, page 2

PLC Programmable logic controller, page 2

Functions

$\hat{\tau}_{up}$ Maximum sojourn time with respect to a current state estimation $\widetilde{X}_t$, page 97

λ Output function of the TAAO, page 23

Λ^{DES} Output function of the DES, page 21

$\lambda_{\|}$ Output function of the $TAAO_{\|}$, page 51

d I/O-subset and I/O-pair distance function, page 71

$edge$ Edge function (to determine the rising and falling edges of an I/O), page 100

$equal$ Equality function (to compare partial output sequences), page 70

ext Tolerance extension function (to extend the determined tolerance interval of a transition), page 35

FD Fault detection policy, page 97

FD_{TDM} Global fault detection policy, page 106

J Language growth criterion of optimal partitioning, page 79

$join$ Join output function (to combine the outputs of two or more partial TAAOs), page 50

$proj_{SUB}^{-1}$ Inverse projection function for a given I/O-subset, page 73

$proj_{SUB,t}$ Timed projection function for a given I/O-subset, page 48

$proj_{SUB}$ Projection function for a given I/O-subset, page 70

$TRes_d^{\cap}$, $TRes_d^{\cup}$ Deadlock behavior residuals, page 101

$TRes_{TDM,d}^{\cap}$, $TRes_{TDM,d}^{\cup}$ Global deadlock behavior residuals, page 107

$TRes_{eb}$, $TRes_{lb}$ Early and late behavior residuals, page 102

$TRes_{TDM,eb}$, $TRes_{TDM,lb}$ Global early and late behavior residuals, page 107

untime Untiming function of a timed word w_t, page 25

vEdge I/O-vector edges (to determine the rising and falling edges between two system outputs), page 101

Operators

min(*sequence*), max(*sequence*) Minimum and maximum value contained in a sequence of natural numbers, page 28

min(*set*), max(*set*) Minimum and maximum value contained in a set of natural numbers, page 27

$|sequence|$ Number of elements contained in a sequence, page 28

$|set|$ Cardinality of a set, page 26

$TAAO \parallel TAAO$ Parallel composition of two partial $TAAOs$, page 51

$vector[i]$ Addresses the i-th element of a vector, page 25

$word\langle a..b\rangle$ Substring operator (to determine a substring in a word between the given positions a and b, page 36

Variables

α Confidence level of the tolerance interval, page 34

β Hit rate of the tolerance interval, page 34

$\perp$ Undefined time span, page 21

δ Output time span, page 21

$\Delta_{EXC}^{(x,x')}$ Exceeding transition time spans related to the transition from state x to state x', page 33

$\Delta_{NR}^{(x,x')}$ Non-reproducible transition time spans related to the transition from state x to state x', page 33

$\Delta_{Obs}^{(x,x')}$ Observed transition time spans related to the transition from state x to state x', page 33

$\Delta_{Orig}^{(x,x')}$ Original transition time spans related to the transition from state x to state x', page 33

Δ_{Mod}^{n} Modeled time span sets for all words of a modeled logical language L_{Mod}^{n}, page 31

$\Delta_{Mod}^{w^q}$ Modeled time spans sets of a modeled logical word w^q, page 31

Δ_{Obs}^{n} Observed time span sequences for all words of an observed logical language L_{Obs}^{n}, page 28

$\Delta_{Obs}^{w^q}$ Observed time span sequences of an observed logical word w^q, page 28

$\Delta_{Orig}^{w^q}$ Original time span sets of an original logical word w^q, page 25

γ Ratio of time-constrained transitions, page 60

Ω Output alphabet of the TAAO, page 23

$\Omega_{\|}$ State outputs of the $TAAO_{\|}$, page 51

Σ Observed output sequences of the DES, page 69

σ_h The h-th observed output sequence in Σ, page 69

Σ_t Observed timed output sequences of the DES, page 22

$\Sigma_{SUB,class}$ Observed partial output sequences related to a class, page 71

$\sigma_{SUB,h}$ The h-th observed partial output sequence in Σ_{SUB}, page 69

$\Sigma_{SUB,t}$ Observed partial timed output sequences of the DES, page 48

$\sigma_{SUB,t}$ Partial timed output sequence of the DES, page 48

Σ_{SUB} Observed partial output sequences of the DES, page 69

$\sigma_{t,h}$ The h-th observed timed output sequence in Σ_t, page 22

τ_{lo}, τ_{up} Lower and upper time bounds of a time guard $[\tau_{lo}, \tau_{up}]$, page 23

θ Generalized time period of the TDM, page 60

$\widetilde{X}_t$, $\widetilde{X}_{t-1}$ Current and former state estimation, page 95

$\widetilde{X}_{i,t}$, $\widetilde{X}_{i,t-1}$ Current and former state estimation related to $TAAO_i$, page 105

$\widetilde{X}_{TDM,t}$ Current state estimation of the TDM, page 105

c Clock of the TAAO, page 23

$C_{\|}$ Clocks of the $TAAO_{\|}$, page 51

CAU Causal I/O-subsets, page 73

D I/O-pair distance matrix, page 73

E^{DES} Events of the DES, page 20

e^{DES} Event of the DES, page 20

ext_0 Tolerance extension parameter, page 34

F Feasible I/O-subsets determined by causal and optimal partitioning, page 84

IO Set of controller inputs and outputs of the DES, page 46

io I/O contained in the set IO, page 46

J_{max} Optimal partitioning parameter, page 79

k Logical identification parameter, page 33

L Logical language of the DES, page 4

$L_{Mod,t}^n$, L_{Mod}^n Modeled timed and modeled logical languages of length n of the DES, page 30

$L^n_{Obs,t}$, L^n_{Obs} Observed timed and observed logical languages of length n of the DES, page 27

$L^n_{Orig,t}$, L^n_{Orig} Original timed and original logical languages of length n of the DES, page 25

l_h Length of the h-th observed timed output sequence $\sigma_{t,h}$ in Σ_t, page 22

L_t Timed language of the DES, page 4

$L_{EXC,t}$ Exceeding timed language of the DES, page 17

$L_{Mod,t}$ Modeled timed language of the DES, page 16

$L_{NR,t}$ Non-reproducible timed language of the DES, page 17

$L_{Obs,t}$ Observed timed language of the DES, page 16

$L_{Orig,t}$ Original timed language of the DES, page 16

m Number of controller inputs and outputs of the DES (i.e. $|IO| = m$), page 46

N Number of I/O-subsets contained in an I/O-partition P (i.e. $|P| = N$), page 46

OPT Optimal I/O-subsets, page 83

P I/O-partition of the DES, page 46

p Number of observed output sequences in Σ_t and Σ, page 22

R^{DES} Transitions of the DES, page 20

SUB I/O-subset of the DES, page 46

t System time of the DES, page 21

t_{SUB} Partial system time related to the I/O-subset SUB, page 48

$TAAO$ Timed Autonomous Automaton with Output, page 23

$TAAO_i$ Partial $TAAO$ defined related to the i-th I/O-subset of a given I/O-partition P, page 49

$TAAO_\|$ Parallel composed $TAAO$, page 51

TDM Timed distributed model, page 49

TG Time guards of the TAAO, page 23

$TG_\|$ Time guards of the $TAAO_\|$, page 51

TT Timed transitions of the TAAO, page 23

$TT_\|$ Timed transitions of the $TAAO_\|$, page 51

u, u_t System output and timed system output of the DES, page 21

$u_{SUB,t}$ Partial timed system output related to the I/O-subset SUB, page 48

u_{SUB} Partial system output related to the I/O-subset SUB, page 47

$u_{TAAO_i,t}$ Partial timed system output with respect to $TAAO_i$, page 49

v Length of a time span sequence or cardinality of a time span set Δ (i.e. $v = |\Delta|$), page 28

v_0 Minimum number of observations required for a tolerance interval with given $1-\alpha$ and β, page 34

w^q Logical word of length q, page 25

$W^q_{Mod,t}$, W^q_{Mod} Modeled timed and modeled logical words of length q of the DES, page 30

$W^q_{Obs,t}$, W^q_{Obs} Observed timed and observed logical words of length q of the DES, page 27

w^q_t Timed word of length q, page 25

X State-space of the TAAO, page 23

X^{DES} State-space of the DES, page 21

x^{DES} State of the DES, page 20

x_0 Initial state of the TAAO, page 23

$x_{0\|}$ Initial state of the $TAAO_{\|}$, page 51

$X_{\|}$ State-space of the $TAAO_{\|}$, page 51

Bibliography

[Aigner, 2007] Aigner, M. (2007). *A Course in Enumeration*. Springer Berlin Heidelberg New York.

[Alur and Dill, 1994] Alur, R. and Dill, D. L. (1994). A theory of timed automata. *Theoretical Computer Science*, 126(2):183–235.

[Basile et al., 2011] Basile, F., Chiacchio, P., Coppola, J., and De Tommasi, G. (2011). Identification of petri nets using timing information. In *3rd International Workshop on Dependable Control of Discrete Systems (DCDS'11)*, pages 154–161.

[Berthomieu and Diaz, 1991] Berthomieu, B. and Diaz, M. (1991). Modeling and verification of time dependent systems using time petri nets. *Software Engineering, IEEE Transactions on*, 17(3):259–273.

[Birolini, 2010] Birolini, A. (2010). *Reliability Engineering*. Springer, 6 edition.

[Blanke et al., 2006] Blanke, M., Kinnaert, M., Lunze, J., Staroswiecki, M., and Schröder, J. (2006). *Diagnosis and Fault-Tolerant Control*. Springer-Verlag New York, Inc., Secaucus, NJ, USA.

[Booth, 1967] Booth, T. (1967). *Sequential Machines and Automata Theory*. John Wiley and Sons, Inc., New-York.

[Bosch Rexroth AG, 2001] Bosch Rexroth AG (2001). Bosch quality training: Mechatronic standard system. Technical report, Robert Bosch GmbH Automation AT-didactic.

[Brandin and Wonham, 1994] Brandin, B. and Wonham, W. (1994). Supervisory control of timed discrete-event systems. *IEEE Transactions on Automatic Control*, 39(2):329–342.

[Cabasino et al., 2007] Cabasino, M. P., Giua, A., and Seatzu, C. (2007). Identification of petri nets from knowledge of their language. *Discrete Event Dynamic Systems*, 17(4):447–474.

[Cassandras and Lafortune, 2008] Cassandras, C. G. and Lafortune, S. (2008). *Introduction to Discrete Event Systems*. SpringerLink Engineering. Springer, 2nd edition.

[Chen and Patton, 1999] Chen, J. and Patton, R. J. (1999). *Robust model-based fault diagnosis for dynamic systems.* Kluwer Academic Publishers, Norwell, MA, USA.

[Choi and Kim, 2002] Choi, S. J. and Kim, T. G. (2002). Identification of discrete event systems using the compound recurrent neural network: Extracting devs from trained network. *Simulation*, 78(2):90–104.

[Cook et al., 2004] Cook, J. E., Du, Z., Liu, C., and Wolf, A. L. (2004). Discovering models of behavior for concurrent workflows. *Computers in Industry*, 53(3):297–319.

[Danancher et al., 2011] Danancher, M., Roth, M., Lesage, J.-J., and Litz, L. (2011). A comparative study of three model-based fdi approaches for discrete event systems. In *3rd International Workshop on Dependable Control of Discrete Systems (DCDS'11)*, pages 29–34.

[Das and Holloway, 2000] Das, S. R. and Holloway, L. E. (2000). Characterizing a confidence space for discrete event timings for fault monitoring using discrete sensing and actuation signals. *IEEE Transactions on Systems, Man and Cybernetics, Part A: Systems and Humans*, 30(1):52–66.

[Dash and Venkatasubramanian, 2000] Dash, S. and Venkatasubramanian, V. (2000). Challenges in the industrial applications of fault diagnostic systems. *Computers and Chemical Engineering*, 24:785–791.

[Debouk et al., 2000] Debouk, R., Lafortune, S., and Teneketzis, D. (2000). A coordinated decentralized protocol for failure diagnosis of discrete event systems. *Journal of Discrete Event Dynamical Systems: Theory and Application*, 10:33–86.

[Dotoli et al., 2008] Dotoli, M., Fanti, M. P., and Mangini, A. M. (2008). Real time identification of discrete event systems using petri nets. *Automatica*, 44(5):1209–1219.

[Dotoli et al., 2009] Dotoli, M., Fanti, M. P., Mangini, A. M., and Ukovich, W. (2009). On-line fault detection in discrete event systems by petri nets and integer linear programming. *Automatica*, 45(11):2665–2672.

[Estrada-Vargas, 2013] Estrada-Vargas, A. P. (2013). *Black-Box identification of automated discrete event systems.* PhD thesis, L'Ecole Normale Superieure de Cachan and CINVESTAV.

[Estrada-Vargas et al., 2010] Estrada-Vargas, A. P., López-Mellado, E., and Lesage, J.-J. (2010). A comparative analysis of recent identification approaches for discrete-event systems. *Mathematical Problems in Engineering*, 2010:21.

[Fanti and Seatzu, 2008] Fanti, M. P. and Seatzu, C. (2008). Fault diagnosis and identification of discrete event systems using petri nets. In *9th International Workshop on Discrete Event Systems (WODES'08)*, pages 432–435.

[Genc and Lafortune, 2003] Genc, S. and Lafortune, S. (2003). Distributed diagnosis of discrete-event systems using petri nets. In van der Aalst, W. and Best, E., editors, *Applications and Theory of Petri Nets 2003*, volume 2679 of *Lecture Notes in Computer Science*, pages 316–336. Springer Berlin Heidelberg.

[Gold, 1978] Gold, E. M. (1978). Complexity of automaton identification from given data. *Information and Control*, 37(3):302–320.

[Grinchtein et al., 2005] Grinchtein, O., Jonsson, B., and Leucker, M. (2005). Inference of timed transition systems. *Electronic Notes in Theoretical Computer Science*, 138(3):87–99.

[Hartung and Klösener, 2009] Hartung, J. a. B. E. and Klösener, K.-H. (2009). *Statistik. Lehr- und Handbuch der angewandten Statistik*. Oldenbourg Wissenschaftsverlag.

[Hashtrudi Zad et al., 2005] Hashtrudi Zad, S., Kwong, R., and Wonham, W. (2005). Fault diagnosis in discrete-event systems: Incorporating timing information. *IEEE Transactions on Automatic Control*, 50(7):1010–1015.

[IEEE Std 100-1996, 1997] IEEE Std 100-1996 (1997). *The IEEE standard dictionary of electrical and electronics terms (IEEE Std 100-1996)*. Institute of Electrical and Electronics Engineers, 6th edition.

[Isermann, 2006] Isermann, R. (2006). *Fault-Diagnosis Systems: An Introduction from Fault Detection to Fault Tolerance*. Springer Berlin.

[ISO/IEC 2382-14:1997, 1997] ISO/IEC 2382-14:1997 (1997). ISO IEC 2382-14: Information technology - Vocabulary - Part 14: Reliability, maintainability and availability.

[Jiroveanu and Boel, 2006] Jiroveanu, G. and Boel, R. (2006). A distributed approach for fault detection and diagnosis based on time petri nets. *Mathematics and Computers in Simulation*, 70(5-6):287–313.

[Kellerer et al., 2004] Kellerer, H., Pferschy, U., and Pisinger, D. (2004). *Knapsack Problems*. Springer Berlin Heidelberg New York.

[Klein, 2005] Klein, S. (2005). *Identification of Discrete Event Systems for Fault Detection Purposes*. Berichte aus der Automatisierungstechnik. Shaker.

[Korte and Vygen, 2012] Korte, B. and Vygen, J. (2012). *Combinatorial Optimization: Theory and Algorithms*. Springer Berlin Heidelberg New York, 5th edition.

[Lefebvre and Leclercq, 2011] Lefebvre, D. and Leclercq, E. (2011). Stochastic petri net identification for the fault detection and isolation of discrete event systems. *IEEE Transactions on Systems, Man and Cybernetics, Part A: Systems and Humans*, 41(2):213–225.

[Lunze, 2012] Lunze, J. (2012). *Automatisierungstechnik*. Oldenbourg Wissenschaftsverlag, 3th edition.

[Maruster et al., 2003] Maruster, L., Weijters, A. J. M. M. T., van den Bosch, A., and Daelemans, W. (2003). Discovering process models by rule set induction. In *Proceedings of the 5th International Workshop on Symbolic and Numeric Algorithms for Scientific Computing*, pages 180–191.

[Meda-Campaña and López-Mellado, 2005] Meda-Campaña, M. E. and López-Mellado, E. (2005). Identification of concurrent discrete event systems using petri nets. In *Proceedings of the of Mathematical Computer, Modeling and Simulation Conference (IMACS'05)*, pages 11–15.

[Medeiros et al., 2007] Medeiros, A., Weijters, A., and Aalst, W. (2007). Genetic process mining: an experimental evaluation. *Data Mining and Knowledge Discovery*, 14(2):245–304.

[Medeiros et al., 2005] Medeiros, A. K. A. D., Weijters, A. J. M. M., and van der Aalst, W. M. P. (2005). Genetic process mining: A basic approach and its challenges. In *Busines Process Management Workshops 2005*, pages 203–215. Springer.

[Narendra and Thathachar, 1974] Narendra, K. and Thathachar, M. (1974). Learning automata - a survey. *IEEE Transactions on Systems, Man and Cybernetics*, SMC-4(4):323–334.

[Nke and Lunze, 2011] Nke, Y. and Lunze, J. (2011). Online control reconfiguration for a faulty manufacturing process. In *3rd International Workshop on Dependable Control of Discrete Systems (DCDS'11)*, pages 19–24.

[Pandalai and Holloway, 2000] Pandalai, D. N. and Holloway, L. E. (2000). Template languages for fault monitoring of timed discrete event processes. *IEEE Transactions on Automatic Control*, 45(5):868–882.

[Papadopoulos and McDermid, 2001] Papadopoulos, Y. and McDermid, J. (2001). Automated safety monitoring: A review and classification of methods. *International Journal of Condition Monitoring and Diagnostic Engineering Management*, 4(4):1–32.

[Philippot et al., 2007] Philippot, A., Sayed Mouchaweh, M., and Carre-Menetrier, V. (2007). Unconditional decentralized structure for the fault diagnosis of discrete event systems. In *1st IFAC Workshop on Dependable Control of Discrete-Event Systems*, pages 67–72.

[Reiter, 1987] Reiter, R. (1987). A theory of diagnosis from first principles. *Artificial Intelligence*, 32(1):57–95.

[Riordan, 2002] Riordan, J. (2002). *Introduction to Combinatorial Analysis*. Dover Publications.

[Rohloff and Lafortune, 2002] Rohloff, K. and Lafortune, S. (2002). On the computational complexity of the verification of modular discrete-event systems. In *Proceedings of the 41st IEEE Conference on Decision and Control*, pages 16–21.

[Roth, 2010] Roth, M. (2010). *Identification and Fault Diagnosis of Industrial Closed-loop Discrete Event Systems*. Logos Verlag Berlin.

[Roth et al., 2010] Roth, M., Lesage, J.-J., and Litz, L. (2010). Identification of discrete event systems: Implementation issues and model completeness. In *Proceedings of the 7th International Conference on Informatics in Control, Automation and Robotics (ICINCO)*, pages 73–80.

[Russell and Norvig, 2010] Russell, S. J. and Norvig, P. (2010). *Artificial Intelligence - A Modern Approach*. Pearson Education, 3rd edition.

[Sampath et al., 1998] Sampath, M., Lafortune, S., and Teneketzis, D. (1998). Active diagnosis of discrete-event systems. *IEEE Transactions on Automatic Control*, 43(7):908–929.

[Sampath et al., 1995] Sampath, M., Sengupta, R., Lafortune, S., Sinnamohideen, K., and Teneketzis, D. (1995). Diagnosability of discrete-event systems. *IEEE Transactions on Automatic Control*, 40(9):1555–1575.

[Sampath et al., 1996] Sampath, M., Sengupta, R., Lafortune, S., Sinnamohideen, K., and Teneketzis, D. (1996). Failure diagnosis using discrete-event models. *IEEE Transactions on Control Systems Technology*, 4(2):105–124.

[Sayed-Mouchaweh, 2012] Sayed-Mouchaweh, M. (2012). Decentralized fault free model approach for fault detection and isolation of discrete event systems. *European Journal of Control*, 18(1):82–93.

[Schneider et al., 2012] Schneider, S., Litz, L., and Lesage, J.-J. (2012). Determination of timed transitions in identified discrete-event models for fault detection. In *51st IEEE Conference on Decision and Control*, pages 5816–5821.

[Supavatanakul et al., 2003] Supavatanakul, P., Falkenberg, C., and Lunze, J. (2003). Identification of timed discrete-event models for diagnosis. In *14th International Workshop on Principles of Diagnosis*, pages 193–198.

[Supavatanakul et al., 2006] Supavatanakul, P., Lunze, J., Puig, V., and Quevedo, J. (2006). Diagnosis of timed automata: Theory and application to the damadics actuator benchmark problem. *Control Engineering Practice*, 14(6):609–619.

[van der Aalst and Weijters, 2004] van der Aalst, W. and Weijters, A. (2004). Process mining: A research agenda. *Computers in Industry*, 53(3):231–244.

[van Schuppen, 2004] van Schuppen, J. H. (2004). System theory for system identification. *Journal of Econometrics*, 118(1-2):313–339.

[Veelenturf, 1978] Veelenturf, L. P. J. (1978). Inference of sequential machines from sample computations. *IEEE Trans. Comput.*, 27(2):167–170.

[Verwer et al., 2006] Verwer, S. E., de Weerdt, M. M., and Witteveen, C. (2006). Identifying an automaton model for timed data. In Saeys, Y., Tsiporkova, E., Baets, B. D., and van de Peer, Y., editors, *Proceedings of the Annual Machine Learning Conference of Belgium and the Netherlands (Benelearn)*, pages 57–64.

[Wang et al., 2007] Wang, Y., Yoo, T.-S., and Lafortune, S. (2007). Diagnosis of discrete event systems using decentralized architectures. *Discrete Event Dynamic Systems*, 17(2):233–263.

[Willner and Heymann, 1991] Willner, Y. and Heymann, M. (1991). Supervisory control of concurrent discrete-event systems. *International Journal of Control*, 54(5):1143–1169.

[Zaytoon and Lafortune, 2013] Zaytoon, J. and Lafortune, S. (2013). Overview of fault diagnosis methods for discrete event systems. *Annual Reviews in Control*, 37(2):308–320.

[Zaytoon and Sayed-Mouchaweh, 2012] Zaytoon, J. and Sayed-Mouchaweh, M. (2012). Discussion on fault diagnosis methods of discrete event systems. In *Proceedings of the 11th International Workshop on Discrete Event Systems (WODES'12)*, pages 9–12.

[Zurawski, 2005] Zurawski, R., editor (2005). *The industrial communication technology handbook*. Industrial information technology series. CRC Press.

About the Author

Stefan Schneider was born on 28th April 1985 in Landstuhl, Germany, and lives currently in Schifferstadt, Germany. He studied electrical/computer engineering at the University of Kaiserslautern focusing on automation technology and automatic control. After his graduation in 2010, Stefan Schneider joined the Institute of Automatic Control at the University of Kaiserslautern under the supervision of Professor Litz. His main field of research was automatic modeling and fault diagnosis of Discrete Event Systems. Since February 2015, Stefan Schneider works as an automation engineer at the company BASF SE in Ludwigshafen, Germany.